THE
IDEA LOGBOOK

....

Dedicated to the Creativity of Inspiration

....

BY TOM RAUSCHER

Design: Gladly done by Graphic Minion Studios
Edited by: Dorian Estanislau and Cathie Kennedy

ISBN: 0-9672613-0-9

This book is printed in the United States of America on recycled paper.

This copy of

THE
IDEA LOGBOOK

Belongs to

_____

_____

_____

Who can be reached at

( ) _____

The Title is

_____

_____

_____

Date

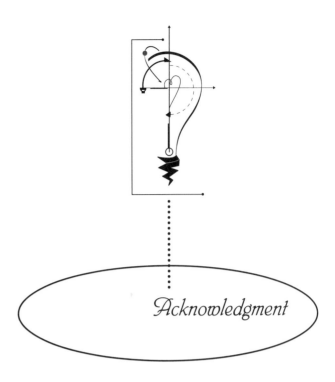

Acknowledgment

This book would not have been possible if it were not for the persistence of many and their constant pleas to share and make available to them, and the world, the lessons learned and shared from my many attempts to develop ideas into products.

It is with great appreciation that I thank "T", Karl, Steve, Chris, Al, H.D. and Monte for their help and support which was warmheartedly given. A special thanks to Randy Mitchell at Awakening Resource Institute for reminding me of my true passion, then assisting and encouraging me to reach my dreams.

Above all, I would like to thank my creator for giving me a tremendous drive, desire, willingness and foresight to trust my instincts and be truly honest with myself.

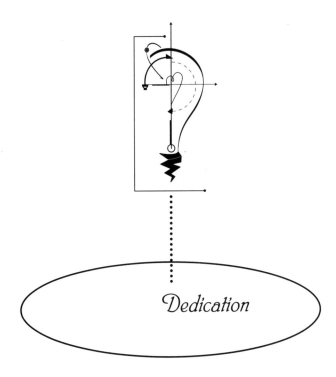

Dedication

The Idea LogBook is dedicated to anyone and everyone who will open their m   i   n   d   s and allow creativity to show them a better way, and then having the fortitude to run with the inspiration.

May inventors reap havoc against today's status quo!

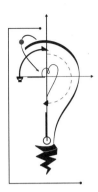

SECTION ONE

SECTION TWO

SECTION THREE

SECTION FOUR

SECTION FIVE

APPENDIX

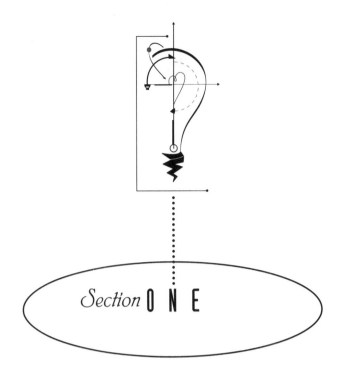

Section O N E

Welcome to the beginning
of your journey.
In this section you will find
information on:

•

the journey itself

•

moving your design forward

•

clarifying your thoughts

•

documenting your thoughts

CONGRATULATIONS

You have just taken the first step in a journey that will bring you tremendous rewards and more satisfaction than you've ever imagined. Your decision to invest in *The Idea LogBook* has brought you one step closer to determining the feasibility of bringing an idea to fruition ... any idea, not just "inventions."

Ideas are a result of your imagination stepping outside "the norm" and looking at something from a completely different perspective. Being open to receiving an inspiration is key to coming up with an idea that's original and marketable. These inspirational ideas can make fortunes.

Upon completion of *The Idea LogBook*, you will have a greater understanding and more clarity of the step-by-step process needed to take your idea from the conceptual phase to production, and then to the market. This *Idea LogBook* will also provide you strategies and insights for protecting your idea, raising finances, and determining the idea's marketability.

A Work in Process

The Idea LogBook is a journal and workbook in one, providing you a realistic process and strong educational base so you can make intelligent and informed decisions and actually enjoy the journey of bringing your inspiration to the market. Its simplistic step-by-step approach gives you down-to-earth guidelines for developing your product while maintaining a high degree of organization.

Only about 1 to 2 percent of new ideas actually make it to the marketplace. Sadly, most people never pursue their ideas. One of the purposes for this book's creation is to raise these statistics through the use of its simple step-by-step guidelines and key information. This working *Idea LogBook* will direct you to conduct preliminary research for your idea and provide valuable information about your product's viability before (if) you spend hundreds and even thousands of hard-earned dollars with companies that promise you the world and deliver nothing.

It is recommended that you first read this journal from cover to cover and then read it once again, while completing the work sections in each chapter. Pay close attention to the "Time for Thought" questions as they help clarify your idea and aspects involved with its development.

There are numerous reasons why individuals have taken on the exciting process of bringing their creation to the market, including: to better the world, make life more simple, and create employment through business start-ups. Some will even admit their real reason for taking this journey is financial gain. That's fine! Wealth is a good thing.

Trivia for Thought

According to the U.S. Department of Health and Welfare, of the "wealthiest of the wealthy," those individuals in the top 1 percent of the wealth category, 1 percent earn it from being in sports, entertainment, winning the lottery or inheritance, 5 percent through sales, 10 percent through service professions (doctors, lawyers, etc.), 10 percent through corporate executive positions and a staggering 74 percent through self-employment!

This process is a self-employed journey.

Upon completion of *The Idea LogBook*, you will be able to make an educated decision on the direction and development of your idea. Each chapter is a step in the educational process to provide you with the knowledge and tools necessary to bring your idea through the various development stages to the marketplace.

Do not concern yourself with whether or not you have all the knowledge, experience or capabilities to take your idea and get it designed and developed. This *Idea LogBook* will provide you with all the information required to do just that. Each area in the development of your idea or product can be accomplished easily.

The Idea LogBook is designed for the development of almost any type of idea. It will allow you to remain organized during this process by following a clear, easy-to-understand system, assisting in all facets of the development.

Within each of the fourteen chapters is a "rule of inspiration." These rules are to provide you with inspiration, encouragement and thoughts to stimulate your own thinking when facing challenging times in your journey. This will help you to keep pushing forward. These rules are all related to the process of product development; but if you should choose to, you can utilize them in any area of your life.

"Appendix A" will have a complete listing of all the "rules of inspiration." Pick a favorite and expand on it or create one or two of your own that inspire you.

☞ RULE # 1 - HAVE FUN.

Kids live by the law of fun and fun alone. It is the sheer Fun Factor of an activity that determines just what a kid is most likely to do. But because we are more mature, we often neglect this need that we still have deep within us. We need to understand the importance of letting the kid inside us come out to play, expressing ourselves more often, exploring and discovering all life's possibilities.

The "Out Door"

If at any phase during this process, for any reason, you discover and realize that this particular idea should just remain an idea, that's fine. In fact, Congratulations! Pat yourself on the back and go out and celebrate.

Celebrate? That's right. You are a winner! You should be commended for what you have already accomplished by taking action. Most people would not have pursued their idea to begin with for fear of failure or a simple lack of ambition. Each time you have at least begun the creation process of developing an idea into product, or for that matter set out on any type of journey, do yourself a great service. Take a few minutes and answer the following questions. They will help clarify the decision you have made.

1) In this process, did you follow your heart?

2) How did it make you feel?

3) Did you take action and obtain results?

4) What have you learned from this process?

....

....

5) What have you discovered about yourself?

....

....

6) How does it feel to have saved hundreds of dollars?

....

....

7) Other thoughts and comments.

....

....

....

....

You will discover and learn some new things about yourself during this process and, no doubt, your character will evolve for the better. Remember that everyone can come up with about four original ideas. The majority never even pursue their inspirations. You are the exception, and you are now classified as an over-achiever.

If you have chosen to use the "out door" for your current idea, that's o.k. Be proud of your accomplishment. The next idea will benefit from the experience you gained and your mind will now clear itself from the thoughts of this idea and be open to receive the next idea(s). Now is the time to order another *Idea LogBook* and get started right away.

Inventors are on the leading edge of creativity and inspiration, traveling an uncertain journey as they put themselves and their idea on the forefront for judgement and scrutiny. The independent inventor built the world. In acknowledgement of this inventor's spirit, *The Idea LogBook* seeks to take the process of invention and make it easy. You can do anything you set your heart to with a step-by-step process. After all, as the saying goes, you can eat an elephant if you take it one bite at a time. That is what this book will help you do. We will take your idea and break down the development process into bite-sized portions so you can accomplish your product goals with ease and precision.

Developing an idea into a product, or any type of creative endeavor, can give you a great sense of achievement, accomplishment and satisfaction , for this do-it-yourself journey will require much of your heart and soul. So again, CONGRATULATIONS on beginning your journey. It will be a journey that you will remember for a lifetime. You now stand on the brink of your inspiration's creation.

Let's have fun.

TIME FOR THOUGHT

There will be a "Time for Thought" section at the end of each chapter. Completing it will increase the overall ease and flow of your journey.

1. What are your reasons for pursuing the development of your idea?

....

....

....

....

2. What do you expect to gain from this journey?

....

....

....

....

3. Describe what would keep you in this process.

....

....

....

....

4. Describe what would interfere with your process.

....

....

....

....

5. Thoughts or comments

....

....

....

....

....

....

....

....

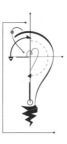

Notes: There will be a "note" section at the end of each chapter and at the very end of this book. This is provided as a place for you to take important notes throughout this journey. Therefore, no additional journal should be required, unless you desire. Be sure to date everything.

NOTES

CONCEPTUALIZATION

Welcome to the beginning of your journey—developing your inspiration.

Take a few minutes and complete the following pages. This will begin the process of taking your idea and clarifying what you have in your mind by stating it in a written format. As you go through this exciting process, be sure to return to this section and add to or revise these answers as your idea evolves. The importance of completing this section, and all sections as described, is to create a trail establishing proof positive that you conceived the idea and continued the journey of developing it.

☞ *RULE # 2 - RIGHT ACTION = RIGHT RESULTS.*

Action - this one simple word has stopped millions of individuals from chasing their (entrepreneurial) dream. A study has determined that this decision is based on a fear, otherwise known as a False Evaluation About Reality. Not knowing what to do, where to turn and how to go about making dreams a reality, most people never get this far off the drawing board. So stop and pat yourself on the back for doing the one thing that is required to make your dreams come true. That is Right **ACTION**.
Make sure you are doing at least three things every day to move you in the direction of your dreams.
Make sure these actions have measurable results you can feel good about.
This process is just one measurable tool to assure you are taking the right action.

Remember this is a "working journal in progress" which means there are no wrong answers. Even if you are not completely comfortable or satisfied with what you initially write, don't be concerned. You should change and revise your journal as you and your idea evolve.

Take your time and enjoy this section. Here are a few guidelines for filling in the upcoming section.

Name/Date - Indicate your name (sign and print) and the date your inspirational journey begins.

Product name - What do you call your idea? Its title? Try to give it a catchy name—something with a zing or ring to it. In most cases, the shorter the name the better.

Description - Describe your idea.

Summary - Explain its advantages and benefits and how it solves a problem(s).

Purpose - What is the intent of your idea? What will it do? Who will use it? Why?

Background - How and why you came up with the idea. Give as much detail as possible as this will be useful to establish the idea's authenticity.

Witness - Go to 2 or 3 of your trusted friends or associates and have them sign and date the form. The purpose if this is twofold. 1) To begin protecting your idea. You will be surprised how valuable this is. You are not telling everyone, just a few close friends or associates who will sign and date the form. This will be discussed more in detail in "Protecting Your Idea" (Chapter 4). 2) By explaining your idea to 2 to 3 others, you will become more comfortable discussing it which could lead you to better clarify the idea. But be OPEN to their reactions (and even suggestions) as you and your idea are evolving.

Once you have completed this section, turn to "Financial" (Chapter 12). Read this section thoroughly as it will assist you in understanding some of the financial obligations you will face during your idea's development. It will also provide an organized structure for tracking your costs through this journey.

RECORD YOUR CONCEPT

I, _____ On _____, _____ begin
        Sign and print name   Date

my (our) journey of developing _____
                                        Name of Product

DESCRIPTION

....
....
....
....
....
....
....
....

SUMMARY

....
....
....
....
....
....
....

PURPOSE

....
....
....
....
....
....
....

R E C O R D   Y O U R   C O N C E P T

B A C K G R O U N D

....

....

....

....

....

....

C O N C E P T U A L I Z A T I O N

_____ _____

Witness/Date

_____

City, State

_____ _____

Witness/Date

_____

City, State

_____ _____

Witness/Date

_____

City, State

Use the following pages to make and record drawings of your idea. These drawings are to be as complete and detailed as you can make them. Be sure to include any dimensions and assign a reference number to each part on the drawing. These reference numbers are required and discussed further when filing the patent applications.

Title each page and date and sign the initial drawing. Date and sign every slightest revision thereafter, even if the revision is drawn right next to your previous drawing. Get your witnesses to also sign the page and any changes.

Once you have completed the description page and drawings of your idea, take this page and have it notarized. This is very important and the first step toward protecting the conception of your idea. This will be explained in more detail in Chapter 4, "Protecting the Idea."

PRODUCT DIAGRAM

Don't worry about your drawing skill level either. Again, you are trying to capture the idea and as much information about it as possible.

There are more of these drawing pages located in the Appendix section should your product require them.

Product Name: _____   Date: _____

Witnessed and understood by: _____ Date: _____

PRODUCT DIAGRAM

Product Name: _____   Date: _____

Witnessed and understood by: _____   Date: _____

PRODUCT DIAGRAM

Product Name: _____   Date: _____

Witnessed and understood by: _____   Date: _____

PRODUCT DIAGRAM

Product Name: _____   Date: _____

Witnessed and understood by: _____   Date: _____

P R O D U C T   D I A G R A M

Product Name: _____   Date: _____

Witnessed and understood by: _____   Date: _____

P R O D U C T D I A G R A M

Product Name: _____   Date: _____

Witnessed and understood by: _____   Date: _____

PRODUCT DIAGRAM

Product Name: _____   Date: _____

Witnessed and understood by: _____   Date: _____

CONCEPTUALIZATION

Product Name: _____   Date: _____

Witnessed and understood by: _____   Date: _____

PRODUCT DIAGRAM

Product Name: _____   Date: _____

Witnessed and understood by: _____   Date: _____

TIME FOR THOUGHT

1. What type of market do you foresee for your idea's usage?

   ....
   ....
   ....
   ....
   ....

2. List any <u>other</u> markets or products that may benefit from your idea.

   ....
   ....
   ....
   ....
   ....

3. What feedback did you get from your witnesses?

   ....
   ....
   ....
   ....
   ....

4. What do you see as the greatest result from your idea?

   ....
   ....
   ....
   ....
   ....

5. What are some of the obstacles that may present themselves to you? How will you overcome them?

   ....
   ....
   ....
   ....
   ....

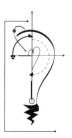

NOTES

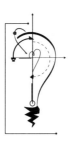

NOTES

CONCEPTUALIZATION

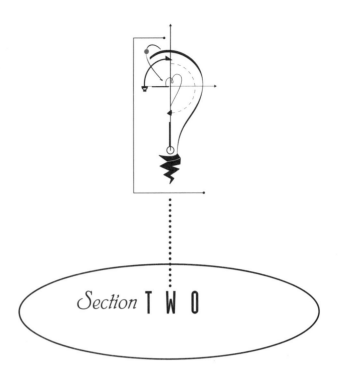

Section **T W O**

In this section we will look at ways
to bring your concept into the three-
dimensional world through the
exciting process of prototyping.

We will also begin to carefully
document your idea and explore
ways to protect your
evolving concept.

PROTOTYPE

Upon completion of the Conceptual stage, you now have a more clear understanding of your idea. It is time to begin developing a prototype. A prototype is an actual model, more or less, of your concept built in three-dimensional space. This is where you take your thoughts and drawings and begin to add form—the muscle that fits on the skeletal frame of your idea. The prototype is usually the most exciting part of the process for an inventor because it is the time when the idea takes shape and an actual entity is created. Through the development of a prototype, a very powerful emotion takes over inventors because they are able to actually put their hands on something tangible that has been created from their inspiration.

This section is the second section (third chapter) of *The Idea LogBook* because it is during this prototype stage that: 1) you are given a boost of excitement and energy from seeing your idea take form as a tangible entity, 2) you can quickly determine whether the idea will work (ie, is it feasible), 3) you can determine if a prototype or product can even be developed, 4) you will begin to understand and determine the uniqueness of your product, and 5) it allows the required time for the product to properly develop and evolve.

☞ RULE # 3 - BE OPEN.

Keeping an open heart and mind allows many doors of possibilities to appear and open.
New ideas come from different, unexpected directions.

A prototype usually refers to an almost completed, ready-for-mass-production and retail sale product where no further refinement is required. However, for simplicity, we will continue to call your product, no matter what stage of its development, a prototype.

Begin the development of your prototype as inexpensively as possible. If you can, create it out of paper, old sheets or cloths, clay, scrap wood, plastic and even discarded metal parts. When building the first prototype, do not use the more costly materials that you will use in the final prototype. In most instances, this first prototype will not perform as you intended anyway. By using "found" materials and inexpensive parts, you will save money, which will be needed more in the later phases of development.

Unexpected situations will surface as the prototype evolves into a product. This is a normal part of the process which actually gives you valuable information to better understand the prototype's flaws. You can determine at this stage whether your idea will function as you expected before spending an excessive amount of money.

Your goal at this point is to begin building this first prototype with the least amount of money (if any). If you do have to spend money, downgrade the materials you choose to the least expensive grade possible.

For example, if you want to make your prototype out of wood, do not buy walnut, maple or even large sheets of plywood, especially if you do not have the tools necessary to work these materials. This will be totally inefficient from a costing standpoint. A better choice of materials is balsa wood which is cheap, extremely easy to work with and very light.

If your prototype is an electrical item, use old broken-down equipment from which to pull parts. Speak with a local repair shop to see if they have any equipment or worn-out parts they are scrapping.

The more you can standardize your product through the use of existing material (standard size bolts, nuts, etc.), the more you will reduce its overall costs, especially in the manufacturing stage of the process, as long as it does not infringe on your actual idea.

Cosmetic Appeal

Do not worry about how this very first prototype looks, because its purpose is to determine if the idea will even function. Be very open at this point, as a completely different shape or idea may evolve.

As you continue through this exciting journey of developing an idea into a product, you will need to refer back to this prototype and your initial drawings. Make changes and revisions as they come to mind. Be sure to date and sign all changes to the drawings.

Upon completion of your first prototype, conduct your own evaluation and determine if it meets your expectations and functions as you intended. Make any adjustments necessary to create the prototype that meets all your requirements, even if it means completely building another prototype. Again, this prototype is built for functionality, not cosmetic appeal.

As you evaluate your prototype and its functions, determine if it could perform any other capabilities or applications other than what you initially intended. Could it have several "spin-offs" that could lead to other products or an overall broader usage? What other ideas surface?

Once the first fully functional prototype is complete and meets all your requirements, it's time to produce a second prototype using more of the material that will eventually be in your finalized prototype. This will determine if the material that you originally expected to use will perform and function with your idea. Some materials have unexpected character changes when combined with other materials or used under certain conditions.

Outside Help

If you are not gifted in production techniques for various prototypes, do not be discouraged. There are product development and prototype companies in existence that will work with you to develop everything you need. Their expertise may also pay off by reducing the time for development of the prototype. To find these development companies, look in your local Yellow pages or ask fellow inventors whom they use for product developing or prototyping. Be sure to interview these companies first to determine if they have the expertise in the industry related to your product and that their fees are within your limits.

You can easily discuss your general prototype requirements with these development companies, but DO NOT go into detail about your idea until they agree to sign a "confidential agreement" protecting you and your idea. See "Protecting the Idea" (Chapter 4) for an explanation and a sample of an agreement form.

As you contact contact these companies or any other company or expert (patent attorney, manufacturer, packaging company, etc.) it is imperative that you make enough contacts so you have a minimum of three that you actually sit down with to discuss your requirements. This will provide you with the knowledge necessary to understand what would be best for your product and assure you are receiving a fair deal.

Utilize the following page to list prototype or product development companies and their qualifications.

PROTOTYPE DEVELOPMENT COMPANIES

PROTOTYPE

Company Name: .... Contact: ....

Telephone: .... Title: ....

Fax: .... E-mail: ....

Company Address: ....

City, State, zip: ....

Qualifications: ....

....

....

Company Name: .... Contact: ....

Telephone: .... Title: ....

Fax: .... E-mail: ....

Company Address: ....

City, State, zip: ....

Qualifications: ....

....

....

Company Name: .... Contact: ....

Telephone: .... Title: ....

Fax: .... E-mail: ....

Company Address: ....

City, State, zip: ....

Qualifications: ....

....

....

Company Name: .... Contact: ....

Telephone: .... Title: ....

Fax: .... E-mail: ....

Company Address: ....

City, State, zip: ....

Qualifications: ....

MATERIALS LIST

The following material lists are to keep a record of your materials, components, part numbers and manufacturers needed to build your prototype. This information is vital in keeping track of your materials and will be a quick reference to procure them for other prototypes or future production. Be sure to go to Chapter 12, "Financial" and indicate the purchase amount on the expense report.

MATERIAL LIST

Product Name _____

Description/Type/Size	Part #	Manufacturer City, State	Quantity

M A T E R I A L S   L I S T CONT.

PROTOTYPE

Product Name _____

Description/Type/Size	Part #	Manufacturer City, State	Quantity

MATERIALS LIST CONT.

Product Name _____

Description/Type/Size	Part #	Manufacturer City, State	Quantity

TIME FOR THOUGHT

If you are not able to answer all of these questions immediately, answer what you can and keep returning to them until they are complete.

1. What would be the least expensive way to build a prototype?

....

....

....

2. What would be the quickest way to build the first prototype?

....

....

....

3. What least-expensive material can you utilize for your first prototype?

....

....

....

4. What material will the final prototype/product be made from?

....

....

....

5. What materials are similar products made from?

....

....

....

6. What other uses could your product fulfill?

....

....

....

TIME FOR THOUGHT CONT.

7. How can the hardware be standardized?

....
....
....

8. What type of assistance or expertise do you require to build your prototypes?

....
....
....

9. What type of service or repair will it require?

....
....
....

10. What type of warranty will you offer?

....
....
....

11. How can the overall quality be improved?

....
....
....

12. How should the product be priced?

....
....
....

13. How does it save money or improve value over the competition?

....
....
....

NOTES

PROTOTYPE

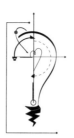

NOTES

PROTECTING THE IDEA

Ssshhhhhh! You have a completely new idea that is so simple it is going to revolutionize its industry. You can even picture yourself on the cover of your favorite magazine and in your brand-new custom-built home, or driving that brand-new luxury import. But if you tell someone about your idea, they will steal it!

Familiar thoughts?

What should you do?

There are several options to take to assure you are protecting your inspiration. As you continue through this process, you will be able to determine the best direction which fits you, your product, and finances.

> ☞ RULE # 4 - IT IS EASY, BECAUSE IT *IS* EASY.

Have you ever had the experience where you are doing something that you think is easy, only to have someone walk up to you and say "Man! How do you do that?! I could never do that like you do!" Probably so. You didn't even think about what you were doing, much less think of it as something difficult. How is this so? The reason is that when you are doing something you are gifted at and love doing, it is second nature—or easy. This is especially true when you are doing what you love to do.

Before spending the thousands of dollars required to process a patent application and receive its acceptance (and that does not include patent attorney fees), it is best to have some basic understanding of what can and cannot be patented so you can make a more informed decision.

Your first step is to contact the United States Patent & Trademark Office (PTO) at 1-800-PTO-9199 or (703) 308-help/4357 and request patent information on the *Disclosure Document, Provisional & Non-Provisional Patents, Trademark* and other *Patent information*. The material is free yet provides valuable information which will provide you with more than a basic knowledge and understanding of the patenting process requirements. Take the time to review this information carefully.

History of Patents

The first patent legislature was enacted in 1790 and has had a few revisions since its inception. Overall, it specifies the subject matter for which a patent may be obtained and the conditions that apply. According to the PTO, only the inventor may apply for the patent; however, it does make considerations to certain circumstances such as a co-inventor, a guardian of the inventor if mentally unfit, the legal executor of the (deceased's) estate, a person having a proprietary interest in the invention, and even if an inventor refuses or cannot be found.

The purpose of obtaining a patent is to assure you of being the originator of the idea, place you in a strong negotiating position should you choose to license it out to someone, and it excludes others

from making, using, offering for sale, or even selling or importing the invention throughout the United States and its territories for a certain length of time.

A general explanation of what can be patented according to the PTO, is *any new discovery and useful process, machine, manufacture, composite of matter and then any new improvements spun off from the* original *idea.* This is why it is important to continue to date and sign all changes and revisions as your idea evolves.

According to the PTO, a *new idea* falls under the umbrella of *Intellectual Property* which consists of:

1. T*rademarks* - a non-generic word, name or symbol, (or combination thereof) and used to identify and distinguish goods from competitors'. The term for a Trademark is 10 years, with 10-year renewal terms; however, it can last indefinitely if the owner continues to use the mark to identify its good and services and completes certain requirements to keep the registration alive. If not, it is cancelled.

2. *Copyrights* - original works of authorship or expressions (movies, books, music, computer programs, etc.). According to the United States Library of Commerce Copyright Office, all works created after January 1, 1978, have copyright protection that will endure for the life of the author plus an additional 70 years.

3. T*rade Secrets* - if you file for a patent and receive a patent pending file number, that number is considered a trade secret, and its legality will depend on whether every appropriate effort is made to maintain strict confidentiality.

4. *Patents* - There are 3 types of patents:

    1. Utility - new or useful processes (chemical, mechanical or electrical) and has a life span of 20 years from file date. The majority of patents fall under this type of patent.

    2. Design - new, original and ornamental design (appearance) for an article of manufacture and has a life span of 14 years from file date.

    3. Plant - asexual reproduction of any distinct and new variety of plants (mutants, hybrids and seedlings) and has a life span of 20 years from file date.

Patents have different filing fees and requirements, so be certain to choose the correct patent to assure proper protection. If a design patent is submitted, it means anyone can make your idea using your concept, they just have to change the design.

New use for a known product should be claimed as a Process Claim and will be objectively examined and a patent granted or not granted based on what the prior patent shows.

Upon receiving a patent application, the PTO will conduct an examination and grant a patent if the new idea is obvious or not obvious in character, it has usefulness, and it has novelty.

For the Record, before you take your idea and show the world or anyone....

The PTO indicates that the day you make your idea known to the public through tradeshows, magazine articles, sales of the product, and even invention development organizations, you have one year to file for a patent application for protection.

Okay, first a reality check. You are a step ahead of the pack for following your inspiration and heart by taking action; however, you might be surprised to discover that many others have come across your idea in the past. Have you ever heard someone say, "Hey, I thought of that years ago and should have done something about it"? Have you ever said something like that?

In most cases, your fear of telling someone about your idea and having them steal it are unwarranted because most people will not do anything for lack of experience and knowledge about developing a product. This does not mean you go around screaming at the top of your lungs to everyone you meet that you have this great idea. It just means you should be selective about whom you tell.

The *Idea LogBook* is also designed to assist in the protection of your idea. Though it does not provide patent protection, it does provide a very powerful documentation of the idea's conception and its evolvement. Proof of conception of an idea is highly regarded by the PTO.

By continuing to date and sign any and all changes and revisions within this *Idea LogBook*, as well as having gathered those 2-3 signatures and obtaining notarization, you have been taking the proper steps to protect your idea's conception. Within the United States, the law indicates that the first person to invent or write the idea is to be rewarded—not the first person to file. This documentation is the first method inventors have to begin legally protecting their idea; in this case, protecting its conception.

Where you must be most cautious in protecting the idea is in the area of the work you have completed by outside assistance or subcontractors such as the printer, packaging, prototype, marketing and developing companies, suppliers, a mentor (unless they are one of your witnesses), etc. Anytime you utilize outside assistance, the simplest and most effective method to protect your idea is to have everyone sign and complete a Confidentiality Agreement. Voila! That's it. You are now protected, and it is very binding proof.

If they will not sign the agreement, ask yourself if you would trust them to babysit your children. This may seem extreme, but you are the one to define the integrity of the people within your surroundings. You may want to also state in the agreement that *all work being completed is proprietary property* and that you hold all the rights to it. For instance, if a packaging company designs a special packaging system or a design artist creates a special logo for your idea, you would want to make sure that you own the *rights* to these developments.

On the following page is a sample of a confidentiality agreement.

Design and develop your agreement to meet your requirements. You may want to obtain proper legal counsel to assure your rights are protected.

CONFIDENTIAL/
NON-DISCLOSURE/NON-COMPETE AGREEMENT

This Agreement is made as of the date of _____ between_____ _____ (hereinafter referred to as EVALUATOR) located at _____ and _____ (hereinafter referred to as INVENTOR) whose address is _____.

This Agreement shall govern the conditions of disclosure between the INVENTOR and the EVALUATOR of certain "Confidential Information" including but not limited to prototypes, drawings, data, trade secrets and intellectual property relating to the "Patent Pending" invention named _____ invented and all rights owned by INVENTOR.

With regard to the Confidential Information the EVALUATOR hereby agrees:

1. Not to use the information therein except for evaluating its interest in entering a business relationship with INVENTOR based on the invention.

2. To safeguard the information against disclosure to others with the same degree of care as exercised with its own information of a similar nature.

3. Not to disclose the information to others, without the express written permission of INVENTOR, except that:

    a. which EVALUATOR can demonstrate by written records was previously known;

    b. which are now, or become in the future, public knowledge other than through acts or omissions of EVALUATOR;

    c. which are lawfully obtained by EVALUATOR from sources independent of INVENTOR.

4. That EVALUATOR shall not directly or indirectly acquire any interest in, or design, create, manufacture, sell or otherwise deal with any item or product, containing, based upon or derived from the information, except as may be expressly agreed to in writing by INVENTOR.

5. That the secrecy obligations of EVALUATOR with respect to the information shall continue for a period ending five (5) years from the date hereof.

INVENTOR will be entitled to obtain an injunction to prevent threatened or continued violation of this Agreement, but failure to enforce this Agreement will not be deemed a waiver of this Agreement.

IN WITNESS WHEREOF the Parties have hereunto executed this Agreement as of the day and year first above written and enforced in accordance with the laws of the State of _____regardless of place of execution or performance.

EVALUATOR _____   INVENTOR _____

By: _____ Date: _____   By: _____ Date: _____

Title:_____   Title:_____

The next method for protecting your idea is to complete a Disclosure Document and send it to the PTO along with the required $10 fee. ***This is not a patent application!*** The disclosure document is only evidence of conception of the idea and will be destroyed unless it is referred to in a related patent application filed within two years. Completing and sending the Disclosure Document application to the PTO adds additional ammunition, along with *The Idea LogBook*, to prove you conceived the idea.

This *Idea LogBook* properly followed meets the Patent's Office's requirement for protecting an idea's conception, but completing the Disclosure Document application and submitting it to the PTO will provide additional, substantial credibility of evidence. Don't bother sending anything to anyone registered mail.

Any time you have a marketing company complete work for you and they promise to file it with the Patent Office as part of their process and fees, they may only complete a Disclosure Document or a Provisional Patent application. The Provisional Patent application takes the Disclosure Document Program one step further in that the PTO not only assigns it a date of conception but also issues a provisional patent pending number. This pending number is vital because it protects your idea **and allows you to sell your product and test the market to determine if it is a moneymaker** before spending thousands of dollars in completing its development.

Let me restate the importance of this. **A Non-Provisional application accepted by the PTO provides you protection for one year, allowing you to generate sales of your developed product.**

The Provisional Patent is good for one year only, and within that year, (not by that year's end) then a Non-Provisional Patent must be filed. The PTO has very strict guidelines regarding the language and content of the Provisional and Non-Provisional Patent applications and will reject them at the slightest inconsistency. The PTO recommends the use of a patent attorney or agent because they are familiar with their format.

The filing fee for a Provisional Application is $75.00 providing you meet the PTO's requirement of being a corporation with under 500 employees; otherwise, the fee is doubled.

The following pages are copies of the Disclosure Document and the Provisional Patent applications and the cover form to be submitted to the PTO. Also, drawings and a detailed description and explanation of your invention must be included with both of these applications. The drawings required and accepted are the same hand drawings you completed in Chapter 2.

In completing the written explanations, the PTO prefers them to be typed, with $1\frac{1}{2}$ inches of line spacing, on regular white 8 $\frac{1}{2}$ by 11 paper with all pages numbered.

Also, be very clear and precise, explaining everything in complete detail. Reference every part number in the description.

If you omit anything, it will be omitted from any protection. Be very detail-oriented.

Be sure to date and sign everything.

Name

Address

City, State

Box DD

Assistant Commissioner for Patents

Washington, DC 20231

Date

To whom it may concern:

The undersigned being the inventor of the disclosed invention called _____, requests that the enclosed papers be accepted under the Disclosure Document Program, and that they be preserved for a period of two years.

Thank you,

Name (print and sign)

PTO/SB/95 (06-1999)
Approved for use through 5/31/2002. OMB 0651-0030
Patent and Trademark Office; U.S. DEPARTMENT OF COMMERCE
Under the Paperwork Reduction Act of 1995, no persons are required to respond to a collection of information unless it displays a valid OMB control number.

Disclosure Document Deposit Request

Mail to:

> **Box DD**
> **Assistant Commissioner for Patents**
> **Washington, DC 20231**

Inventor(s):_____

Title of Invention:_____

Enclosed is a disclosure of the above-titled invention consisting of _____ sheets of description and _____ sheets of drawings. A check or money order in the amount of _____ is enclosed to cover the fee (37 CFR 1.21(c)).

The undersigned, being a named inventor of the disclosed invention, requests that the enclosed papers be accepted under the Disclosure Document Program, and that they be preserved for a period of two years.

_____   _____
Signature of Inventor   Address

_____   _____
Typed or printed name

_____   _____
Date   City, State, Zip

NOTICE TO INVENTORS

It should be clearly understood that a Disclosure Document is not a patent application, nor will its receipt date in any way become the effective filing date of a later filed patent application. A Disclosure Document may be relied upon only as evidence of conception of an invention and a patent application should be diligently filed if patent protection is desired.

Your Disclosure Document will be retained for two years after the date it was received by the Patent and Trademark Office (PTO) and will be destroyed thereafter unless it is referred to in a related patent application filed within the two-year period. The Disclosure Document may be referred to by way of a letter of transmittal in a new patent application or by a separate letter filed in a pending application. Unless it is desired to have the PTO retain the Disclosure Document beyond the two-year period, it is not required that it be referred to in the patent application.

The two-year retention period should not be considered to be a "grace period" during which the inventor can wait to file his/her patent application without possible loss of benefits. It must be recognized that in establishing priority of invention an affidavit or testimony referring to a Disclosure Document must usually also establish diligence in completing the invention or in filing the patent application since the filing of the Disclosure Document.

If you are not familiar with what is considered to be "diligence in completing the invention" or "reduction to practice" under the patent law or if you have other questions about patent matters, you are advised to consult with an attorney or agent registered to practice before the PTO. The publication, *Attorneys and Agents Registered to Practice Before the United States Patent and Trademark Office*, is available from the **Superintendent of Documents, Washington, DC 20402**. Patent attorneys and agents are also listed in the telephone directory of most major cities. Also, many large cities have associations of patent attorneys which may be consulted.

You are also reminded that any public use or sale in the United States or publication of your invention anywhere in the world more than one year prior to the filing of a patent application on that invention will prohibit the granting of a patent on it.

Disclosures of inventions which have been understood and witnessed by persons and/or notarized are other examples of evidence which may also be used to establish priority.

There is a nationwide network of Patent and Trademark Depository Libraries (PTDLs), which have collections of patents and patent-related reference materials available to the public, including automated access to PTO databases Publications such as *General Information Concerning Patents* are available at the PTDLs, as well as the PTO's Web site at www.uspto.gov. To find out the location of the PTDL closest to you, please consult the complete listing of all PTDLs that appears on the PTO's Web site or in every issue of the Official Gazette, or call the PTO's General Information Services at 800-PTO-9199 (800-786-9199) or 703-308-HELP (703-308-4357). To ensure assistance from a PTDL staff member, you may wish to contact a PTDL prior to visiting to learn about its collections, services, and hours.

Burden Hour Statement: This collection of information is used by the public to file (and by the PTO to process) Disclosure Document Deposit Requests. Confidentiality is governed by 35 USC 122 and 37 CFR 1.14. This collection is estimated to take 12 minutes to complete, including gathering, preparing, and submitting the completed Disclosure Document Deposit Request to the PTO. Time will vary depending upon the individual case. Any comments on the amount of time you require to complete this form and/or suggestions for reducing this burden, should be sent to the Chief Information Officer, U.S. Patent and Trademark Office, U.S. Department of Commerce, Washington, D.C., 20231. DO NOT SEND FEES OR COMPLETED FORMS TO THIS ADDRESS. SEND TO: Assistant Commissioner for Patents, Washington, D.C. 20231.

PROTECTING THE IDEA

PROTECTING THE IDEA

Please type a plus sign (+) inside this box → ☐

PTO/SB/16 (2-98)
Approved for use through 01/31/2001. OMB 0651-0037
Patent and Trademark Office; U.S. DEPARTMENT OF COMMERCE
Under the Paperwork Reduction Act of 1995, no persons are required to respond to a collection of information unless it displays a valid OMB control number.

✚

PROVISIONAL APPLICATION FOR PATENT COVER SHEET
This is a request for filing a PROVISIONAL APPLICATION FOR PATENT under 37 CFR 1.53 (c).

INVENTOR(S)

Given Name (first and middle [if any])	Family Name or Surname	Residence (City and either State or Foreign Country)

☐ *Additional inventors are being named on the ___ separately numbered sheets attached hereto*

TITLE OF THE INVENTION (280 characters max)

CORRESPONDENCE ADDRESS

Direct all correspondence to:

☐ Customer Number | [Type Customer Number here] → | *Place Customer Number Bar Code Label here*

OR

☐ Firm *or* Individual Name |
Address |
Address |

| City | | State | | ZIP | |
| Country | | Telephone | | Fax | |

ENCLOSED APPLICATION PARTS (check all that apply)

☐ Specification *Number of Pages* [ ]   ☐ Small Entity Statement

☐ Drawing(s) *Number of Sheets* [ ]   ☐ Other (specify) [ ]

METHOD OF PAYMENT OF FILING FEES FOR THIS PROVISIONAL APPLICATION FOR PATENT (check one)

☐ A check or money order is enclosed to cover the filing fees

☐ The Commissioner is hereby authorized to charge filing fees or credit any overpayment to Deposit Account Number: [ ]

FILING FEE AMOUNT ($)
[ ]

The invention was made by an agency of the United States Government or under a contract with an agency of the United States Government.

☐ No.

☐ Yes, the name of the U.S. Government agency and the Government contract number are: _____

_____

Respectfully submitted,

SIGNATURE _____

TYPED or PRINTED NAME _____

TELEPHONE _____

Date [/ /]

REGISTRATION NO.
(if appropriate) [ ]
Docket Number: [ ]

USE ONLY FOR FILING A PROVISIONAL APPLICATION FOR PATENT

This collection of information is required by 37 CFR 1.51. The information is used by the public to file (and by the PTO to process) a provisional application. Confidentiality is governed by 35 U.S.C. 122 and 37 CFR 1.14. This collection is estimated to take 8 hours to complete, including gathering, preparing, and submitting the complete provisional application to the PTO. Time will vary depending upon the individual case. Any comments on the amount of time you require to complete this form and/or suggestions for reducing this burden, should be sent to the Chief Information Officer, U.S. Patent and Trademark Office, U.S. Department of Commerce, Washington, D.C., 20231. DO NOT SEND FEES OR COMPLETED FORMS TO THIS ADDRESS. SEND TO: Box Provisional Application, Assistant Commissioner for Patents, Washington, D.C., 20231.

✚

PTO/SB/09 (12-97)
Approved for use through 9/30/00. OMB 0651-0031
Patent and Trademark Office; U.S. DEPARTMENT OF COMMERCE
Under the Paperwork Reduction Act of 1995, no persons are required to respond to a collection of information unless it displays a valid OMB control number.

STATEMENT CLAIMING SMALL ENTITY STATUS (37 CFR 1.9(f) & 1.27(b))—INDEPENDENT INVENTOR	Docket Number (Optional)

Applicant, Patentee, or Identifier: _____

Application or Patent No.: _____

Filed or Issued: _____

Title: _____

As a below named inventor, I hereby state that I qualify as an independent inventor as defined in 37 CFR 1.9(c) for purposes of paying reduced fees to the Patent and Trademark Office described in:

☐ the specification filed herewith with title as listed above.

☐ the application identified above.

☐ the patent identified above.

I have not assigned, granted, conveyed, or licensed, and am under no obligation under contract or law to assign, grant, convey, or license, any rights in the invention to any person who would not qualify as an independent inventor under 37 CFR 1.9(c) if that person had made the invention, or to any concern which would not qualify as a small business concern under 37 CFR 1.9(d) or a nonprofit organization under 37 CFR 1.9(e).

Each person, concern, or organization to which I have assigned, granted, conveyed, or licensed or am under an obligation under contract or law to assign, grant, convey, or license any rights in the invention is listed below:

☐ No such person, concern, or organization exists.

☐ Each such person, concern, or organization is listed below.

Separate statements are required from each named person, concern, or organization having rights to the invention stating their status as small entities. (37 CFR 1.27)

I acknowledge the duty to file, in this application or patent, notification of any change in status resulting in loss of entitlement to small entity status prior to paying, or at the time of paying, the earliest of the issue fee or any maintenance fee due after the date on which status as a small entity is no longer appropriate. (37 CFR 1.28(b))

_____	_____	_____
NAME OF INVENTOR	NAME OF INVENTOR	NAME OF INVENTOR
_____	_____	_____
Signature of inventor	Signature of inventor	Signature of inventor
_____	_____	_____
Date	Date	Date

Burden Hour Statement: This form is estimated to take 0.2 hours to complete. Time will vary depending upon the needs of the individual case. Any comments on the amount of time you are required to complete this form should be sent to the Chief Information Officer, Patent and Trademark Office, Washington, DC 20231. DO NOT SEND FEES OR COMPLETED FORMS TO THIS ADDRESS. SEND TO: Assistant Commissioner for Patents, Washington, DC 20231.

PROTECTING THE IDEA

Non-Provisional Patent Application

The final method for protecting your idea is to file for an actual Non-Provisional Patent application to the PTO. This patent provides you the most thorough protection. The processing fee for a Non-Provisional application is $380 for a small corporation (doubled if over 500 employees). If the application is accepted, costs will reach about $4,000 which will include numerous other charges such as maintenance, processing, etc.

However, prior to filing for a patent, you may want to conduct a patent search. These searches determine whether a patent currently exists or if the idea has been written in a publication. <u>If a search results in the discovery of a prior patent, written permission from the holder of the patent must be obtained</u> so as not to incur a lawsuit.

Patent searches are conducted in the Patent and Trademark Office in Arlington, Virginia, one of the Patent and Trademark Depository Libraries and even on the Internet. These libraries have various capabilities and functions, and a list of their locations should be included in the package sent to you from the PTO. A quick and easy patent search can be conducted over the Internet at the PTO website at <u>www.uspto.gov</u> or <u>ftp.uspto.gov</u>; however, this listing does not include the work in process and any that has yet to be submitted in their site.

Another place to conduct a search is at IBM's Intellectual Property site which allows you to conduct a free search at <u>www.patents.IBM.com</u>. It is designed to search through the United States patents back to 1971 and contains over two million patents that can be retrieved along with a wealth of other valuable information.

Another site that is becoming a powerful place to gather product development information and even sell your new idea is <u>www.inventorsplace.com</u>. The IBM, PTO and other informative sites can be accessed from it.

If you have access to the Internet, go to these sites regularly to spend some time. They provide a wealth of information worthy of your invested time and will keep you informed of product development happenings.

Should I Do a Patent Search?

Though it is not required by law to conduct a patent search prior to filing for a patent application, doing a search could save you money, time, and effort in continuing with a dead-end endeavor. If a patent is discovered, you must cease immediately from this endeavor and obtain the legal right to continue with your invention or begin developing your next idea.

Filing, Filing, Filing ... Must I Do This?

Filing for the patent application can be a gruesome task or an exciting challenge. If you ask anyone who has ever filed for one, most would agree it is gruesome. There are many strict requirements to file for a patent, but they are designed to ensure some type of conformity and fairness to all. Because of the complexity and legality, no detailed information is included about filing for a Non-Provisional Patent application. If you choose to complete the application yourself, it is recommended that you follow the guide located in the information package you requested from the PTO. However, if further assistance is required, the PTO has technical individuals on staff whom you may contact.

Once the application has been filed at the PTO, the patent takes about 12 to 18 months to process before a patent is received but until then, a pending number is issued. Once you have your patent pending number, you are now in a stronger position to begin negotiating with manufacturers, marketers and venture capitalists (should you choose) to begin mass development of your idea. The patent pending number is considered a trade secret.

Trademarks

Depending on the name or title you give your product, you may choose to have it trademarked. The PTO's definition of a *trademark* is a phrase, symbol or design or a combination of words, phrases, symbols or designs which identifies and distinguishes the source of the goods or services of one party from those of others. The PTO trademark application fee is $245.00 with renewable fees of $300.00 and its life span is 10 years to indefinitely, depending on the completion of certain requirements. Again, the PTO recommends you obtain the services of a qualified patent attorney who is familiar with their requirements to complete the form with ease.

In choosing to trademark your product, it is highly advisable to conduct a search. If you have decided to complete a search and file for a patent and are not comfortable with or do not choose to spend the time and energy required, it is recommended to hire a patent attorney or agent to complete the task. Trademark searches are completed just like any other patent.

Do I Need an Attorney?

The PTO looks more favorably on applications received written from patent agents and attorneys because they are familiar with the requirements, language and content. Patent attorneys and agents are qualified to provide advice and perform all the requirements for processing and filing an application. Though the agent's fees are less, the advantage of an attorney is they can provide legal advice and represent you in court if an infringement occurs.

Just for the record, many patent attorneys use agents to complete the application and they review the work to ensure it meets their satisfaction.

Contact and interview several attorneys and/or agents to determine and understand what they will do for you, what experience they have, if they are knowledgeable in the industry of your product's specific field, and if they are listed on the Patent Office register—especially have all fees thoroughly explained.

Most of the time a client/attorney privilege applies to protect your idea, but be certain to ask. Sometimes the attitude in their response will indicate the type of person you would be dealing with.

To obtain names of patent attorneys or agents, look in the Yellow pages of your local phone directory, ask any fellow inventors for their recommendations, contact other attorneys and/or go to the Patent Office web site at www.uspto.gov or ftp.uspto.gov and search their listing. Again, this site has a wealth of patent information that is worthy of your visit.

Utilize the following pages. They will keep you organized as you contact patent attorneys and agents to determine which is best for your product.

Once you receive your patent or trademark number issued by the PTO, it is only valid in the United States and its provinces. In order to protect your rights in foreign countries, you must file in each of these countries to assure proper protection. This is another area you must be certain your patent agency can handle, should it be important to you.

P A T E N T   A T T O R N E Y S   O R   A G E N T S

name _____   address _____

phone _____   referred by _____

date initial contact _____

qualifications_____

    years in practice _____   registered with PTO _____

    industry of expertise _____   request info package _____

    date/time of meeting _____

    referrals of satisfied clients

1. _____   contacted_____

2. _____   contacted_____

3. _____   contacted_____

Fees/other information

_____

_____

name _____   address _____

phone _____   referred by _____

date initial contact _____

qualifications_____

    years in practice _____   registered with PTO _____

    industry of expertise _____   request info package _____

    date/time of meeting _____

    referrals of satisfied clients

1. _____   contacted_____

2. _____   contacted_____

3. _____   contacted_____

Fees/other information

_____

_____

name _____   address _____

phone _____   referred by _____

date initial contact _____

qualifications_____

    years in practice _____   registered with PTO _____

    industry of expertise _____   request info package _____

    date/time of meeting _____

    referrals of satisfied clients

1. _____   contacted_____

2. _____   contacted_____

3. _____   contacted_____

Fees/other information

_____

_____

name _____   address _____

phone _____   referred by _____

date initial contact _____

qualifications_____

    years in practice _____   registered with PTO _____

    industry of expertise _____   request info package _____

    date/time of meeting _____

    referrals of satisfied clients

1. _____   contacted_____

2. _____   contacted_____

3. _____   contacted_____

Fees/other information

_____

_____

- -

name _____   address _____

phone _____   referred by _____

date initial contact _____

qualifications_____

    years in practice _____   registered with PTO _____

    industry of expertise _____   request info package _____

    date/time of meeting _____

    referrals of satisfied clients

1. _____   contacted_____

2. _____   contacted_____

3. _____   contacted_____

Fees/other information

_____

_____

- -

name _____   address _____

phone _____   referred by _____

date initial contact _____

qualifications_____

    years in practice _____   registered with PTO _____

    industry of expertise _____   request info package _____

    date/time of meeting _____

    referrals of satisfied clients

1. _____   contacted_____

2. _____   contacted_____

3. _____   contacted_____

Fees/other information

_____

_____

PATENT ATTORNEYS OR AGENTS CONT.

name _____   address _____
phone _____   referred by _____
date initial contact _____
qualifications_____
    years in practice _____   registered with PTO _____
    industry of expertise _____   request info package _____
    date/time of meeting _____
    referrals of satisfied clients

1. _____   contacted_____
2. _____   contacted_____
3. _____   contacted_____

Fees/other information

_____
_____

name _____   address _____
phone _____   referred by _____
date initial contact _____
qualifications_____
    years in practice _____   registered with PTO _____
    industry of expertise _____   request info package _____
    date/time of meeting _____
    referrals of satisfied clients

1. _____   contacted_____
2. _____   contacted_____
3. _____   contacted_____

Fees/other information

_____
_____

name _____   address _____
phone _____   referred by _____
date initial contact _____
qualifications_____
    years in practice _____   registered with PTO _____
    industry of expertise _____   request info package _____
    date/time of meeting _____
    referrals of satisfied clients

1. _____   contacted_____
2. _____   contacted_____
3. _____   contacted_____

Fees/other information

_____
_____

The timeframe for completing the various applications and getting them to the PTO depends on where your product is in its development and marketing stages. For instance: if your idea requires extensive research and a long leadtime to develop, complete the Disclosure Document application. This gives you two years of protection on the idea's conception. Then, within two years, complete the Provisional application, providing your idea one more year of protection in which you can go out and market it—generating sales and determining if it's a moneymaker. If your product proves successful in the market, then submit the Non-Provisional application providing it optimal patent protection.

OR: If your idea can (will) be developed and ready for market within a couple of months, just complete the Provisional application, thus giving you one year to market your product, generate sales and determine if it is a moneymaker. If it is successful, then submit the Non-Provisional application providing it optimal patent protection.

However, if you feel it is in the best interest of your product, you can choose to submit the Non-Provisional application at any time, thus eliminating the various avenues available and providing it the optimal patent protection.

PROTECTION CHECKLIST

1. _____ This *LogBook* signed by two or three associates and then notarized.

2. _____ Patent office called (1-800-PTO-9199) and patent information, Disclosure Document, Provisional, Non-Provisional and Trademark information and application forms requested. Date completed _____.

3. _____ Searched PTO's and IBM's websites.

4. _____ Patent attorneys and agents contacted and interviewed.

5. _____ Patent search conducted.

6. _____ Disclosure Document prepared.

7. _____ Drawings and detailed description and explanation complete.

8. _____ Date Disclosure Document mailed to PTO _____.

9. _____ Provisional application prepared.

10. _____ Drawings and detailed description and explanation complete.

11. _____ Provisional Application and documentation mailed to PTO.

12. _____ PTO assigned Provisional number _____

13. _____ Non-Provisional Patent application and proper documentation prepared.

14. _____ Non-Provisional Patent application and documentation submitted to PTO.

15. _____ Trademark search and application completed and mailed (optional).

16. _____ Patent Pending number _____

Time for Thought

1. Where do you expect to use Confidential Agreements?

    ....

    ....

    ....

2. What kind of feedback are you getting from the confidential agreements from your required outside assistance?

    ....

    ....

    ....

3. Describe your time frame for processing the

    _____ Disclosure Document

    _____ Provisional Application

    _____ Patent Application

4. What are some reasons your idea would require the use of a patent attorney or agent?

    ....

    ....

    ....

5. Describe what you have learned from contacting and interviewing patent attorneys and agents.

    ....

    ....

    ....

6. After reading the requested information from the PTO and spending some time on their and IBM's site, describe other concerns about protecting your idea. How do you feel you can overcome these concerns?

    ....

    ....

    ....

    ....

    ....

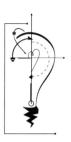

NOTES

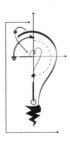

NOTES

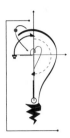

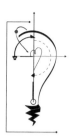

NOTES

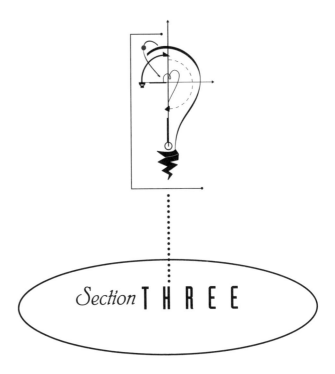

Section **T H R E E**

In this section we will look at ways
to continue the evolution of your
product idea.

•

You will test and refine your idea,
exploring ways to gather marketing
information.

•

We will also take a look at
what is involved in manufacturing
your product.

MARKET EVALUATION

This section focuses on determining whether or not your idea is a viable product. *Viable?* In other words, does a market really exist for your product, and if so, where is it?

Most people are able to come up with an average of four ideas that they feel would lead to a viable product that could better the world and provide them a tremendous financial income. Of all these possible products, though, only a little over 100,000 people annually actually take some type of action. The rest of the people, sadly, do nothing. So by coming this far in the *LogBook*, you are an exception and should commend yourself.

Many others who do pursue their ideas invest hundreds (those are the fortunate ones) or even thousands of dollars with companies who promise to conduct all the necessary work to bring their ideas to the market. These are the companies that promise to provide a steady stream of royalties so you can *live happily ever after.* In most cases, the only thing that happens is that hard-earned money is sunken into a bottomless hole, never providing a return to these would-be inventors.

Conducting a market evaluation is a crucial step in determining whether a market for a product exists. This can be accomplished easily with just a little time and effort, saving you a tremendous amount of money.

☛ RULE # 5 - FOCUS ON PROFITS.

In order for your idea to succeed, it must make a profit for everyone. Develop a "win–win" attitude in assuring that you, your manufacturers and your distributors each make a profit.
Make sure your customers feel they are getting high quality and value.

What Else Is Out There?

Part one in the evaluation process is to better understand the market: Who are your product's competitors and what have they been doing to market their products? This is a very simple task that can be completed within a few hours.

Go to the "Time for Thought" section in this chapter and re-read the questions as they will help you be more aware of the expected requirements.

The first step can be completed as you are out conducting your normal day-to-day activities, or you may want to make a special trip to focus solely on this step. Go to the mall or specialty stores in your local area that carry products similar to your idea or your idea's market. You might be surprised at the many "woozles" that are in close proximity to your idea. Look at these products and notice how they are packaged, the materials they are made from, the colors used, how are they displayed, and their overall appearance.

On the following pages write the company information including name, address and phone number of the manufacturers or any information about the company displayed on the packaging. This information is usually located somewhere on the back or very bottom of the packaging. (At this time do not worry about indicating the Standard Industrial Classification/North American Industrial Classification System (SIC/NAICS) number, sales, or employees spaces. This will be completed at a later time.)

SHOPPING SPREE

MANUFACTURERS OF SIMILAR PRODUCTS

Mfg. 1 _____

_____

_____

_____

SIC   Sales   Emp.

Mfg. 2 _____

_____

_____

_____

SIC   Sales   Emp.

Mfg. 3 _____

_____

_____

_____

SIC   Sales   Emp.

Mfg. 4 _____

_____

_____

_____

SIC   Sales   Emp.

Mfg. 5 _____

_____

_____

_____

SIC   Sales   Emp.

Mfg. 6 _____

_____

_____

_____

SIC   Sales   Emp.

Mfg. 7 _____

_____

_____

_____

SIC   Sales   Emp.

Mfg. 8 _____

_____

_____

_____

SIC   Sales   Emp.

Mfg. 9 _____

_____

_____

_____

SIC   Sales   Emp.

Mfg. 10 _____

_____

_____

_____

SIC   Sales   Emp.

Mfg. 11 _____

_____

_____

_____

SIC   Sales   Emp.

Mfg. 12 _____

_____

_____

_____

SIC   Sales   Emp.

Shopping Spree cont.

Mfg. 13 _____

_____

_____

_____

SIC   Sales   Emp.

Mfg. 14 _____

_____

_____

_____

SIC   Sales   Emp.

Mfg. 15 _____

_____

_____

_____

SIC   Sales   Emp.

Mfg. 16 _____

_____

_____

_____

SIC   Sales   Emp.

Mfg. 17 _____

_____

_____

_____

SIC   Sales   Emp.

Mfg. 18 _____

_____

_____

_____

SIC   Sales   Emp.

Mfg. 19 _____

_____

_____

_____

SIC   Sales   Emp.

Mfg. 20 _____

_____

_____

_____

SIC   Sales   Emp.

Mfg. 21 _____

_____

_____

_____

SIC   Sales   Emp.

Mfg. 22 _____

_____

_____

_____

SIC   Sales   Emp.

Mfg. 23 _____

_____

_____

_____

SIC   Sales   Emp.

Mfg. 24 _____

_____

_____

_____

SIC   Sales   Emp.

PART TWO: Research, Research, Research....

The second part of your market evaluation requires going to the local library (the branch with a good business section). If you are not familiar with your library, this would be a great time to become its friend. The library will offer you a tremendous amount of information that is instrumental to the development of your idea. And guess what - IT IS FREE, FREE, FREE!

Be especially friendly and polite to the librarians behind the reference desk. They will provide a wealth of assistance beyond your expectations, so get to know them. If you do not have a library card or it has expired, now would be a good time to get set up with a new one. If you have access to the Internet, you should be able to conduct most of this same research there. This gives you the added advantages of being in the comfort of your home and being able to do your research at a time that is most convenient for you.

Where Do I Look?

Begin by looking in the many reference books and business listings for the names of those companies that you gathered from your shopping spree. Also, every library has computers with manufacturing and business disks which provide the information at your fingertips.

Below is a list of some of the resources that are available. It is not a complete list because individual states and counties may have additional reference material specific to the businesses in their areas. They are not listed in any specific order and the ones with an * will be ones you will most definitely use.

- Chamber of Commerce Business Directories
- Encyclopedia of Associates
- Business Directory, Fabricators, Assemblers*
- Manufacturing Register*
- Directory of Wholesalers
- Technology Directories
- Bio-Medical
- Commerce and Industry Directory
- Local Business Referral Directory
- U.S. Census Report*
- American Business Directory*
- Brands and their Companies
- Companies and their Brands
- Directory of Corporate Affiliations
- Directory of U.S. Corporations
- Dun's Business Rankings
- Dun & Bradstreet's Annual Million Dollar Directory*
- S&P Register of Corporations, Directors & Executives*
- Thomas Register of American Manufacturers *
- Thomas Food Industry Register
- World Aviation Directory

You should be able to find most of the companies on your list by utilizing one of these sources, but go on if you cannot.

Take each company you listed on page 65 and within the spaces provided, including the surrounding area and margins, write their sales, number of employees, phone numbers and the names of the President and Vice Presidents (make sure the spelling is accurate) and verify their address. Also indicate the listed SIC/NAICS numbers. The SIC/NAICS numbers were created by the government as a system to classify companies by their type(s) of business. They are cross-referenced in almost all the available sources.

These numbers are key in locating other companies in your idea's related industry.

Now you can search companies by these SIC/NAICS number(s), either with the library computer or through the numerous books. Again if you need assistance, you now should know your friendly librarian by name. Write, print or copy the additional list of the companies with these same SIC/NAICS numbers that fit your product's market.

The following pages can be utilized as a place to list these additional companies and keep them organized. If more pages are needed, just make additional copies. These companies that produce similar products to your idea are very important should you decide to sell or license the idea. This is the method and list (to begin with) of companies to contact.

COMPANIES IN SIMILAR INDUSTRIES

Company Name
Point of Contact & Title
Address
Telephone   Fax
SIC/NAICS   Web Address
Notes

Company Name
Point of Contact & Title
Address
Telephone   Fax
SIC/NAICS   Web Address
Notes

Company Name
Point of Contact & Title
Address
Telephone   Fax
SIC/NAICS   Web Address
Notes

COMPANIES IN SIMILAR INDUSTRIES CONT.

Company Name
Point of Contact & Title
Address
Telephone   Fax
SIC/NAICS   Web Address
Notes

Company Name
Point of Contact & Title
Address
Telephone   Fax
SIC/NAICS   Web Address
Notes

Company Name
Point of Contact & Title
Address
Telephone   Fax
SIC/NAICS   Web Address
Notes

Company Name
Point of Contact & Title
Address
Telephone   Fax
SIC/NAICS   Web Address
Notes

Company Name
Point of Contact & Title
Address
Telephone   Fax
SIC/NAICS   Web Address
Notes

COMPANIES IN SIMILAR INDUSTRIES CONT.

Company Name

Point of Contact & Title

Address

Telephone   Fax

SIC/NAICS   Web Address

Notes

Company Name

Point of Contact & Title

Address

Telephone   Fax

SIC/NAICS   Web Address

Notes

Company Name

Point of Contact & Title

Address

Telephone   Fax

SIC/NAICS   Web Address

Notes

Company Name

Point of Contact & Title

Address

Telephone   Fax

SIC/NAICS   Web Address

Notes

Company Name

Point of Contact & Title

Address

Telephone   Fax

SIC/NAICS   Web Address

Notes

COMPANIES IN SIMILAR INDUSTRIES CONT.

Company Name
Point of Contact & Title
Address
Telephone   Fax
SIC/NAICS   Web Address
Notes

Company Name
Point of Contact & Title
Address
Telephone   Fax
SIC/NAICS   Web Address
Notes

Company Name
Point of Contact & Title
Address
Telephone   Fax
SIC/NAICS   Web Address
Notes

Company Name
Point of Contact & Title
Address
Telephone   Fax
SIC/NAICS   Web Address
Notes

Company Name
Point of Contact & Title
Address
Telephone   Fax
SIC/NAICS   Web Address
Notes

Company Name
Point of Contact & Title
Address
Telephone   Fax
SIC/NAICS   Web Address
Notes

Company Name
Point of Contact & Title
Address
Telephone   Fax
SIC/NAICS   Web Address
Notes

Company Name
Point of Contact & Title
Address
Telephone   Fax
SIC/NAICS   Web Address
Notes

Company Name
Point of Contact & Title
Address
Telephone   Fax
SIC/NAICS   Web Address
Notes

Company Name
Point of Contact & Title
Address
Telephone   Fax
SIC/NAICS   Web Address
Notes

COMPANIES IN SIMILAR INDUSTRIES CONT.

Company Name
Point of Contact & Title
Address
Telephone   Fax
SIC/NAICS   Web Address
Notes

Company Name
Point of Contact & Title
Address
Telephone   Fax
SIC/NAICS   Web Address
Notes

Company Name
Point of Contact & Title
Address
Telephone   Fax
SIC/NAICS   Web Address
Notes

Company Name
Point of Contact & Title
Address
Telephone   Fax
SIC/NAICS   Web Address
Notes

Company Name
Point of Contact & Title
Address
Telephone   Fax
SIC/NAICS   Web Address
Notes

Company Name
Point of Contact & Title
Address
Telephone   Fax
SIC/NAICS   Web Address
Notes

Company Name
Point of Contact & Title
Address
Telephone   Fax
SIC/NAICS   Web Address
Notes

Company Name
Point of Contact & Title
Address
Telephone   Fax
SIC/NAICS   Web Address
Notes

Company Name
Point of Contact & Title
Address
Telephone   Fax
SIC/NAICS   Web Address
Notes

Company Name
Point of Contact & Title
Address
Telephone   Fax
SIC/NAICS   Web Address
Notes

COMPANIES IN SIMILAR INDUSTRIES CONT.

Company Name
Point of Contact & Title
Address
Telephone   Fax
SIC/NAICS   Web Address
Notes

Company Name
Point of Contact & Title
Address
Telephone   Fax
SIC/NAICS   Web Address
Notes

Company Name
Point of Contact & Title
Address
Telephone   Fax
SIC/NAICS   Web Address
Notes

Company Name
Point of Contact & Title
Address
Telephone   Fax
SIC/NAICS   Web Address
Notes

Company Name
Point of Contact & Title
Address
Telephone   Fax
SIC/NAICS   Web Address
Notes

Company Name
Point of Contact & Title
Address
Telephone   Fax
SIC/NAICS   Web Address
Notes

Company Name
Point of Contact & Title
Address
Telephone   Fax
SIC/NAICS   Web Address
Notes

Company Name
Point of Contact & Title
Address
Telephone   Fax
SIC/NAICS   Web Address
Notes

Company Name
Point of Contact & Title
Address
Telephone   Fax
SIC/NAICS   Web Address
Notes

Company Name
Point of Contact & Title
Address
Telephone   Fax
SIC/NAICS   Web Address
Notes

Once you've completed the previous section, it's time to learn what the sales market is like and what products are making news within your product's related markets.

Finding the Market News

While still at the library, go back to the computer you used to retrieve all the previous information, or go to the magazine and newspaper sections (again, if uncertain where to find these resources, ask your librarian). Look up topics related to your idea's market by listing some of the names of the larger companies. Spend time reading articles about the top stories, the market, what companies are doing, who is doing what and where the market is currently. Take note of its growth direction as related to whether it will be growing or declining in the coming years. Get an understanding of the size of the industry, the key players, government regulations and even positive and negative happenings. Take notes in the "note" section of this chapter.

Explore the many other areas that present themselves as you conduct this search.

Other valuable sources that provide a wealth of information are available to you free of charge.

U.S. Census Report

Every ten years the U.S. government conducts a survey to obtain an abstract view of the United States population, and their findings are available in most libraries.

Standard and Poor's Industry Surveys

S&P provides in-depth insight and statistics concerning different areas of most industries.

Since you now have a new or renewed library card, go to the marketing section and browse through the books. Bring home and read one that provides you additional knowledge to better assist in evaluating the market for your product idea.

These steps you have just conducted usually cost between 400 and 600 dollars for someone else to complete. You are now way ahead of the game, and you can be certain that your investment into *The Idea LogBook* has *more than paid for itself.*

You have just conducted research. Yes, research! You may be well on your way to becoming a research junkie!

TIME FOR THOUGHT

1. Describe the size of your product's market.

   ....

   ....

2. What is the potential of the market?

   ....

   ....

   ....

3. What are the characteristics of the consumers?

   ....

   ....

4. Who are the major competitors/players?   What are their annual sales?

   a. _____   _____

   b. _____   _____

   c. _____   _____

   d. _____   _____

   e. _____   _____

5. Where are they located?

   a. _____   _____

   b. _____   _____

   c. _____   _____

   d. _____   _____

   e. _____   _____

6. What changes in the market have occurred in the last

   a. 5 years?

   ....

   ....

   b. 10 years?

   ....

   ....

   c. 20 years?

   ....

   ....

TIME FOR THOUGHT CONT.

7. What trend is expected in

a. 5 years?

....

....

b. 10 years?

....

....

....

c. 20 years?

....

....

....

8. What is the cost range of similar products?

a. from _____ to _____

b. from _____ to _____

c. from _____ to _____

d. from _____ to _____

e. from _____ to _____

9. How has the market evaluation effected your prototype development?

....

....

....

10. Does the market have sufficient potential? Explain.

....

....

....

....

11. What are your overall thoughts and concerns about the market?

....

....

....

....

....

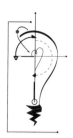

NOTES

MARKET EVALUATION

NOTES

PRODUCT EVALUATION

Thinking of an idea is the easiest part of inventing, and from that point on is when the real challenge and excitement begins. As your prototype becomes more refined, so will its overall evaluation as to its position in the marketplace.

The product evaluation phase is where the invention is objectively evaluated to determine its character, strength, and, yes, weaknesses which will help you better understand its potential in the market.

This chapter was intentionally created as a separate chapter from the "Prototype" section in order to give your prototype the needed time to evolve. This has also provided you with the necessary time to contemplate in greater detail the numerous possibilities that will arise from your product. As the prototype develops, so may its functions, advantages and benefits.

☞ *RULE # 6 - BE WILLING TO LEARN & TAKE RISKS.*

The more you challenge yourself, the more you will leave your comfort zone, and things that seem risky today will become easy tomorrow. Nothing ventured, nothing gained. No risk, no glory.

Is This Rocket Science?

No. It is still an easy, step-by-step process. Before beginning your work in this section, understand that you will be conducting a *SUBJECTIVE,* not *objective* evaluation. This is because there is no one who can predict, with 100 percent accuracy, the success of any product. This concept can be easily experienced. The next time you are at your local grocery market, notice the different items that are new this week or the items no longer on the shelf. Buyers understand that people determine what will stay and what will go by the sales of an item.

Even though the marketing evaluation is based on opinion or best guess, it will provide important strength to your product should you choose to license or sell the idea. This will be discussed in Chapter 10, "Licensing." If you are uncomfortable conducting this part of the process, there are companies and even universities that you can contact to perform this task. However, if you maintain a high degree of integrity and remove all (okay *most*) emotional attachment to the success of the idea, you should be able to easily complete this section. Furthermore, if you have been following the process of this journal, you have already completed some of the requirements. Save your money and just press on. You are doing great!

The biggest advantage of outsourcing the product evaluation process is that the company completing it has no emotional attachment to its success.

What About a Technical Evaluation?

The technical evaluation is a test to determine if your product performs exactly as you state it does. Most patent items do not require such an evaluation unless they are mechanical or electrical in nature.

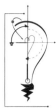

Does your product truly perform as you are stating? Truly?

If you choose to get a technical evaluation completed elsewhere, contact technical schools, universities and even state and local agencies in your area. Only work with a firm in the industry that is related to your product. Remember that you are still going to want to work with people you feel you can trust. Don't forget to use Confidential Agreements.

The best technical evaluation you will ever receive will be FREE and conducted by the company to which you sell the idea. You can be certain that this evaluation will be extremely complete and thorough, assuring the product functions properly.

It is also highly recommended that you go to the library and check out a book on product evaluations for a further, more in-depth understanding of this subject.

Proceed to the following page and answer the questions. Put some time and thought into these answers, even if it takes a few different sittings to complete.

TIME FOR THOUGHT

1. What is the product's market niche?

   ....

   ....

2. Where is the immediate market?

   ....

   ....

3. Why will someone purchase your product?

   ....

   ....

4. List 10 features of your product (i.e., outstanding quality).

   1. _____   6. _____
   2. _____   7. _____
   3. _____   8. _____
   4. _____   9. _____
   5. _____   10._____

5. List 10 benefits of your product (advantages).

   1. _____   6. _____
   2. _____   7. _____
   3. _____   8. _____
   4. _____   9. _____
   5. _____   10._____

6. Describe how it fits the consumer needs.

   ....

   ....

   ....

7. How do you want a manufacturer to think of your product?

   ....

   ....

   ....

8. How do you want a licensing company to think of your product?

   ....

   ....

   ....

TIME FOR THOUGHT CONT.

9. How do you want consumers to think of your product?

    ....

    ....

    ....

10. What will the expected operating costs be?

    ....

    ....

    ....

11. What type of warranty will be needed?

    ....

    ....

    ....

12. What is its expected length of operation?

    ....

    ....

    ....

13. What is the expected wholesale price?

    ....

    ....

    ....

14. What is the expected retail price?

    ....

    ....

    ....

15. List the 5 top competitive products.

    1. _____   4. _____
    2. _____   5. _____
    3. _____

16. List the 5 top competitive companies (including sales).

    1. _____   4. _____
    2. _____   5. _____
    3. _____

17. List the advantages of your product over the competitors.

....
....
....
....

18. What are other industrial uses for your product?

....
....
....
....

19. What savings will it create?

....
....
....
....
....

20. List resources required for its development (suppliers).

....
....
....
....
....

21. Why might your product NOT work?

....
....
....
....

22. What can you do to make it work more efficiently?

....
....
....
....
....

NOTES

PRODUCT EVALUATION

NOTES

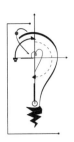

NOTES

NOTES

PRODUCT EVALUATION

MARKETING

Now that you've become a research junkie, it's time to determine if your product will sell.

Yes, everyone knows you would buy it and use it every single day. And though this may be true, it does not amount to a hill of beans. A cold hard fact of reality—if you are looking to create a business or sell it to someone to mass-produce, *your* opinion of *your* idea means very little in determining whether it will be successful.

Harsh as that may sound, accept it now and just reprogram your mindset to focus on the **consumer** who will actually purchase and use your product.

The best resource that is readily available for determining whether your idea will sell is this same **end user** or **consumer**. THEY may be very unpredictable, but THEY are the experts in knowing what THEY want. The more you focus on satisfying *their* needs and wants, the better chance your idea has of becoming successful in the marketplace or becoming commercialized.

Continually ask yourself, "What would be the end-users' perception?" Does the idea actually satisfy a need and want? How does it satisfy a need and want? Think back to when you were browsing the various products in the stores. What was your opinion of these products?

It will be worth your while to go back to the stores and look at the various items from this new and important perspective.

☛ *RULE # 7 - HAVE FAITH IN YOURSELF.*

You can do anything when you put your mind to it. Your behavior must follow your belief. Consistently remind yourself of all your successes and accomplishments.

Determining if Someone Will Buy Your Idea

Okay, now it's time to have some fun and determine if someone will even purchase your idea. Time for marketing. Here are several approaches, which you may utilize at your discretion.

Each method will only be providing a subjective evaluation and by no means will the results indicate exactly how your idea will do in the marketplace as indicated earlier. There is no one person or method that can predict precisely the success of any product. The purpose of marketing your idea is to increase the chance of its success and, more importantly, it provides you, the inventor, a tremendous amount of ammunition to develop a better product.

Survey Says ...

The first option is to conduct a survey. A survey is a list of questions that you have created that will be given to individuals to answer. This will provide you with a wealth of information about the type of individuals who might purchase your product. Relax! This is very easy, simple, and fun to do whether or not you are a mathematician.

Do not allow any family or friends to take the survey as they are too biased toward you. Though that is a good thing for you personally, it will not provide the needed information that will best serve your product's interest.

It is very important to focus your survey toward the buying group of your product to obtain more meaningful results that will better determine the product's potential. For instance, a manufacturer would not care for the opinions of a single mom with four kids for its 2-seater sports car. Also, whatever market your product is in, it is best to focus on that industry. For instance, if your idea is related to the medical or auto industry, get individuals who are in that field to take the survey.

Where Do I Find These People?

There are numerous approaches to get this survey into the hands of individuals. One method is to stand out in front of the local post office, library or grocery store (be sure to get permission first) and politely ask individuals (leaving) to assist you by taking a market survey.

Be totally honest and up front with your potential survey participants. Inform everyone that you are an inventor and are in need of their help. Politely inform them it will take only 2 to 3 minutes of their time. Smile, hand them the clipboard with the survey and ask for their help. Dress above casual, maintain a very bright and positive attitude and always smile. Thank everyone, whether they help you or not.

If your potential survey participants have their hands full and they still would enjoy helping you, you could ask them the questions and complete the form for them. Be extra sensitive to the amount of time that you are taking up with these people. If you find yourself in this situation, you will note that some of the questions can easily be completed without asking the individual. Questions like age range, male/female, etc. It is their honest and sincere opinions and thoughts that are important, not who actually fills out the survey.

As an incentive and only as a suggestion, you may choose to offer them a couple of dollars for their time. You will be surprised how money influences people. By giving 50 people a couple dollars for their honest and sincere opinions, and it saves or makes you thousands, everyone wins.
Hopefully people will just be happy to help you.

Do I Have to Do This Live and in Person?

The greatest advantage of conducting a survey face-to-face with individuals is that it allows you to choose the individuals whom you want to survey. For instance, if your idea is a new purse design, there would be no need to survey men (most do not carry a purse). If your product is a toy, you can focus on children, but be sure to ask their parents if they would let their child help you.

Check's in the Mail ... or at Least Your Survey Is.

Another survey option is to conduct a mailing. Grab your local phone directory and pull out names and addresses of people and mail them your survey with a cover letter politely explaining who you are and what you want. Be sure to include a self-addressed, stamped envelope, or you will not be likely to receive anything back from your mailing.

Mailers from the phone directory provide some ability to focus toward males or females; however, there is no way of knowing who actually completes the survey.

A more accurate and focused mailing list can be obtained by returning to the library and asking your friendly librarian, whom you probably owe lunch by now, for the Standard & Poor's Industry Survey reference book (sometimes these manuals are located behind their desk). Look up the industry or industries related to your idea. You will be able to purchase (very reasonably) a mailing list that focuses directly on your product's field of interest. These lists are usually sold in groups of 1000 names. You can mail out to twenty-five at a time until you reach a desired amount in return, which should be at least fifty for a starting basis.

The greatest disadvantage of mailers is that only about 2 percent of your addressees will even respond. However, to increase this percentage, maybe offer a money-back "thank you." Indicate on the cover that if they complete the survey and mail it back within a week, they will receive X number of dollars (no more than 5 dollars). If you choose to offer money, it should increase the percentage of completed surveys you receive back. Make sure you send the money promised so as not to hinder your integrity or place you on the wrong side of legality; it's the right thing to do.

Following is a list of sample questions to aid in the development of your own market survey. Construct your survey of questions based on your product and its benefits and advantages. These questions are not related to every type of product nor do you need a survey with this many questions. Keep the number of survey questions limited to about 12 to 15. Since people will be helping you, it would be irritating to them if it took them more than 2 to 3 minutes to complete. Also, these questions are not in any particular order and may need to be adjusted to fit your product. For example, you may require a shorter age range starting at 18-24 or 25-30.

Turn the page to examine this sample survey and then start your own marketing masterpiece.

Sample Marketing Survey

Hello, my name is _____.

I am a beginning inventor and am conducting a simple survey about _____ (subject of idea).

Your open and candid views are very much appreciated. You will note that this is an anonymous survey so your information will remain confidential. I have enclosed a stamped return envelope for your convenience.

Thank you in advance for your help.

This portion would be added if you were using this form as a mailer. Obviously, you would not say this if you were conducting the survey in person.

Please place an X or check by the appropriate response.

1. Gender

_____ Female   _____ Male

2. Age

____18-25   ____45-54

____26-34   ____55-64

____35-44   ____over 64

3. Marital status

___ Single (never married)

___ Married

___ Divorced/Separated

___ Widow/Widower

4. Highest academic degree earned

___ H.S.   ___ B.A./B.S.   ___ Ph.D./Ed.D.

___ A.A.   ___ M.A./M.D.

5. Occupational status

___ Professional

___ Managerial/Executive

___ Administrative/Clerical

___ Engineering/Technical

___ Marketing/Sales

___ Skilled craft or Trade

___ Semiskilled occupation

6. Total household income (please check one)

___ $15,000 - $24,000   ___$45,000 - $54,000

___ $25,000 - $34,000   ___ $55,000 and over

___ $35,000 - $44,000

7. Number of people in your household _____

8. Do you own a home (includes townhome, condominium or co-op) as a primary residence?
___ Yes ___ No

9. Do you own additional home(s)?
___ Yes ___ No

10. How many miles (average) do you travel per day (work, store, etc.)? ____

11. Would you consider purchasing a product that offers (list two or three benefits such as more comfortable, easier to clean, improved simplicity or service, increased safety)?
___ Yes ___ No

12. If you answered yes to question 11, what would you be willing to pay?
$_____

13. How often do you currently use (related products or industry)
___ Once a day ___ Once a year
___ Once a month ___ Never

14. How important is (name a benefit) ?
___ very ___ never noticed
___ somewhat ___ not interested

15. How important is (name an advantage)?
___ very ___ never noticed
___ somewhat ___ not interested

16. What benefit(s) would you like to see?
_____
_____

17. What benefits are not important?
_____
_____

18. What price range would you be willing to pay?
from _____ to _____

Drawing Conclusions From Your Survey

After you have obtained satisfactory feedback from your survey, it is time to determine what it all means. Begin by taking a blank survey form and compiling the answers from each question onto one form.

For example, if you used question 1 from above as your first question and you have fifty responses, your analysis should look like this.

1. Gender

___29___ Female   ___21___ Male

Now take these numbers and determine their percentages.
For example:
1. Gender

___29/58%___ Female   ___21/42%___ Male

(29 ÷ 50 = 58% and 21 ÷ by 50 = 42%)

50 = the total number of people surveyed.

Go ahead and do this evaluation for every question.

Now comes the part of taking the tallied survey results and putting them in a written form. Utilizing the survey and its results, answer the following questions by simply stating your results in sentence form.

S U R V E Y   S U M M A R Y :

Take all the tallied information and write a brief explanation describing the results of each question in the survey.

_____
_____
_____
_____
_____
_____
_____
_____
_____
_____
_____
_____
_____
_____
_____
_____
_____
_____

Non-Statistical Results and Discussion

Describe anything that might have happened while conducting the survey that is not related to the actual survey itself. (peoples' reaction about the product, survey, etc.):

_____

_____

_____

_____

_____

_____

_____

_____

_____

_____

_____

_____

MARKETING

Conclusion

The survey not only obtained information about the people but also about specific areas relating to your idea. Describe the overall conclusion and include the tallied percentages from the survey questions. Now just simply describe what you feel it means in just a few sentences.

_____

_____

_____

_____

_____

_____

_____

_____

_____

_____

Conducting a "Hands-On" Survey

Another approach to determine if someone will purchase your product is to actually place your prototype in the hands of people and get feedback from their thoughts and opinions. Being able to utilize as many senses as possible from individuals will provide a more accurate measurement of their feelings about a product.

A place to take your idea and place it in the hands of such individuals to conduct this type of survey (or even sell your product) is a tradeshow. Actual sales of your product are the best form of feedback you can get. There are numerous shows in which you can lease a booth and display your idea for sale. You could even conduct a quick survey. There is a list of tradeshows you can obtain in the *Tradeshows and Professional Exhibits Directory* at your new favorite place, the library.

However, be certain you know and fully understand the promoters of tradeshows. Carefully read the contract because these promoters and their contracts vary. Some have been known to indicate in the signed agreement that they own the rights to a percentage of your idea and will receive a percentage of its sales. Read the contract thoroughly, and since there are many tradeshows, only play with the ones that best fit you and your idea.

Caution! Displaying Your Product Will Affect Your Chances of Obtaining a Patent.

The law states that once an idea is made public for show or sale, the patent application must be filed within one year. See "Protecting the Idea " (Chapter 4) for more details.

As mentioned earlier, actual sales are the best feedback for determining if your product will sell, but a more cost-effective way to determine this, depending on you and your product, is to place a few products on the shelves of your local store or shop to see if anyone purchases them.

This will be next to impossible at most of the giant retail stores, but some of the smaller specialty stores and shops may go along with it for a share of the profits.

Can I Really Sell This Thing? Getting Your Product in a Store.

Go to a small retail outlet, preferably one owned by someone with whom you have established a relationship through the patronage of their store, and ask for the owner. Inform them who you are, what you are proposing, and ask them if you may have a small area of shelf space to display your product. Do not tell or show them your product until they are first open to the concept. Just explain the concept and the type of industry to which your product is related. If you are not that friendly with any store owners, perhaps now is the time to start. Remember, if they balk at the idea, you can simply go to another store. If there is difficulty at this stage, you will have even more difficulty collecting your money. You decide on the quality of individuals with whom you associate.

Inform them you would like to display a small number of your products (let them decide on the amount) for which you require no money up front. You will wait to be paid as the products are sold. Inform the store owner the minimum price for which the products need to be sold and how much money they will get for each item sold. If none sell, you will be happy to remove them from their place of business. This is consignment marketing for new products.

If the owner decides to sell them for more than your minimum price, let them. It is their store and they know the market better. There is no harm in having a higher price for your product if it will sell at that price. Besides, if and when they do sell for that higher price, you now know a better price that the market will pay for your product. This is very powerful once you are ready to introduce your product on a mass market scale.

Let the store owner know that you will stop by once a week to conduct an inventory, determine their satisfaction with the product's movement, replenish or remove the products upon their discretion and exchange moneys. Also, offer a 30-day full money-back guarantee to them and their customers. Whether you are close to the owner or not, it would be wise to draw up a contract explaining the agreement, assuring everyone understands what is expected from this agreement. Make sure your arrangement is fair to all parties.

Again, you may want to contact a lawyer for proper advice to ensure you are not stepping over the boundaries of the law.

Be certain the contract at least includes:
- your objectives
- your name & address
- name of the product
- name of store and owner
- the minimum selling price
- how profits will be divided

- when money will be exchanged
- product guarantees
- and the duration of the contract.

Make sure the owner has your phone number and can easily reach you. That way if a situation out of the ordinary should arise, it can be quickly solved. This is especially important if there is a high demand for your product. It gives them a way to contact you to order more product.

Will a Store Owner Go for This?

The owner of the store or shop has nothing to loose and everything to gain by putting your product on their shelves. The advantage for the owner is that they do not have any money to invest in inventory, from the purchasing of a product and from the risk of not selling anything and then being stuck with product. If your product does not sell, they just simply return it to you. Owners will only stand to make money. This part they will like.

The greatest concern owners might have is if your product is in direct conflict with their current suppliers. In actuality, this is really not an issue since you have invented a completely new and unique product - right?

Two other possible concerns for a store or shop are product liability and dissatisfied customers. In reality, selling your product should have no bearing on their existing liability insurance, and your 30-day guarantee states that you stand behind your product, assuring total customer satisfaction.

If a circumstance arises and a customer *does* return your product to the store, happily and immediately give the owner back the money that was paid to the customer, less the amount already received from the sale of it. Quickly pick up the product and take it with you. Ask the owner if the customer's reason for returning it is known. This will provide you valuable product analysis.

Shop owners are business people and are in business to make profits; otherwise, they would be out of business. If your product does not conflict with their current products and it makes them a nice profit with no cash investment, they should be more than happy to allow your product on their shelves.

How Can I Get the Word Out About My New Product Placement?

Once the store owner has agreed to "carry" your product, be sure to tell everyone and anyone you know the name and location of where your product is for sale. Ask if they would go see it. This promotes your product and the store that was kind enough to display it.

Do not be concerned about whether any of your friends purchase your product or not. Just get them to the store to see it. Perhaps they just might purchase something else in the store. Now you have just added a tremendous amount of value to your product in the eyes of that owner because word of mouth is the best advertisement. The owner will remember everything that happened from this experience and, if all went smoothly, will be happy to assist you with your next idea.

Another great advantage to putting your product on the shelf of a store is that the money you will make from the sale of it can then be used to pay for the patent process or further improvements or development.

Again, as a reminder, displaying your product will affect your chances of obtaining a patent. The law states that once an idea is made public for show or sale, the patent application must be filed within one year.

Please see "Protecting the Idea" (Chapter 4) for inexpensive methods of protecting your idea.

Create Drawings, Create Interest & Generate Money

Another very powerful technique for marketing your idea is to create drawings, create interest and generate money (CD&I-GM) up front before investing thousands of dollars into your idea's development. This approach could have been inserted into the "Financial" chapter because of its ability to save thousands of dollars; however, its strength is its emphasis more on the marketing aspect of product development. The strategy behind this focuses on obtaining drawings and then conducting a marketing campaign to wholesale companies to determine their interest in purchasing your product.

The first step is to create very high quality, glossy, professional drawings of your product combined with text information about your product otherwise known as brochures. These brochures can be only one page on one side but have a clear picture of your product with a list of its advantages and benefits. The proper graphic artist will be able to place everything on the page in a very organized and professional manner. If you do not have the capability to produce these drawings, contact graphic artists in your area and utilize their services.

Graphic artists can be found in the telephone directory, colleges or universities, but the best method to locate them is to ask around. Someone knows someone who can easily create these brochures. The fees for their services vary as well as their skills and abilities, so be sure they meet your requirements.

Use the following as a guideline in contacting graphic artists or designers.

CONTACTING GRAPHIC DESIGNERS

1. Name of designer _____ phone _____
background/qualifications _____
_____
fees/notes _____
_____
_____

2. Name of designer _____ phone _____
background/qualifications _____
_____
fees/notes _____
_____
_____

3. Name of designer _____ phone _____
background/qualifications _____
_____
fees/notes _____
_____
_____

4. Name of designer _____ phone _____
background/qualifications _____
_____
fees/notes _____
_____

CONTACTING GRAPHIC DESIGNERS CONT.

MARKETING

5. Name of designer _____ phone _____

background/qualifications _____

_____

fees/notes _____

_____

_____

6. Name of designer _____ phone _____

background/qualifications _____

_____

fees/notes _____

_____

_____

7. Name of designer _____ phone _____

background/qualifications _____

_____

fees/notes _____

_____

_____

8. Name of designer _____ phone _____

background/qualifications _____

_____

fees/notes _____

_____

_____

9. Name of designer _____ phone _____

background/qualifications _____

_____

fees/notes _____

_____

_____

10. Name of designer _____ phone _____

background/qualifications _____

_____

fees/notes _____

_____

_____

Contact Companies That May Resell Your Product

While your drawings are being completed, it is time to locate companies that may be interested in buying your product to resell. You will want to find the person responsible for purchasing new products at these companies. You will be looking for companies at a wholesale and retail level, which means they will take your product and then turn around and sell it for additional profits.

There are hundreds of thousands of companies that you can contact. The best method for locating them is to return to the library to the computer databases (as you have already done) and look up some of the largest companies that first come to your mind. From there, take note of the SIC/NAICS numbers that closely match your product. Then you can look up companies by this number. You could even drive around a few blocks in your local town or go to the mall.

There are also manufacturers who are looking for new products to add to their existing product mix. For a listing of manufacturers to contact, see the Thomas Register directory at the library or on the Internet.

You should now have a list of several hundred companies that you can begin contacting. Before contacting the extremely large companies, allow yourself some practice by contacting a couple of the smaller ones. Contact these companies and ask for the person in charge of adding new products to their product line. Get the proper pronunciation and spelling of this person's name. Ask them if you may make an appointment to show them a new product; or, if they are not close, ask if you may send a package of a new product for their consideration.

The Review Process: What to Know

If you are able to go to meet with these companies in person, bring several of your brochures and sit down and explain your product, its advantages and benefits and the potential profits they can expect. If you have a prototype in perfect functioning order and it looks very similar to the picture, bring it and let the buyer examine it.

If you cannot go in person to present your product when initially speaking to the buyer, inform them that you will be sending product information for their review. After you have sent the product information (brochure, cover letter and profit potential break down) give the company several days to review it. Then follow up with a phone call to the buyer.

Be prepared to answer questions regarding how the product is to be packaged and shipped, the product's pricing, number in a case, minimum case order, shelf life (if applicable) and terms. Inform them that your product is in the pre-manufacturing phase and they can have it delivered by _____ date (give yourself ample time here) from time of order.

Be Ready With Pricing Information

It is important to be ready to discuss the pricing and profit potential of the product you are presenting. You will want to include a cost break-down sheet that explains their costs for the item, its suggested retail price and the potential profits. Cost this out on a "per piece" basis and also provide a section showing the profits possible from a substantial amount of product sold. If you tier the pricing structure, be certain to include this breakdown and profit potential for each price.

You may choose to tier your pricing to provide incentives to wholesalers to purchase larger quantities of product. The higher the quantity purchased, the lower their cost should be. The advantage is that they can make more profit and you can move more product.

As you are discussing your product with them, listen for any feedback. It could be valuable in your product's development. Ask them if they feel this product would complement their product line. Then ask if they would like to place an order. If they say yes, ask if they will generate a purchase order.

The Purchase Order

The purchase order will almost guarantee the interested company will purchase X quantities of your product. No money is exchanged until delivery of the product or other predetermined terms. However, they could cancel the purchase order any time at will prior to product being delivered. Regardless, this purchase order is *proof positive* that your product has market potential and indicates favorable outlook that its further development, including manufacturing, should be a good investment.

At this point in the process, should you need finances for completing the development (especially manufacturing), these numerous purchase orders for your product are like having concrete collateral.

Begin Your Marketing Campaign

Use the following as a guideline for contacting wholesale companies.

While you are conducting this marketing campaign, if you have not already determined who will manufacture the product, go to Chapter 8. Immediately begin this process, because if you do get orders for your product, you will need to deliver as soon as possible. Buyers for these companies will want to know when to expect delivery so they can plan their selling strategy. You do not want these buyers to wait too long because they may lose their interest in even carrying your product.

Remember, the idea behind this technique is to determine interest in your product. This will provide a wealth of information to you to coordinate the best number of products to begin manufacturing at the least possible cost while achieving the maximum quantities required.

CD&I-GM COMPANIES

1. Name of Company _____ point of contact _____
address _____ phone _____
qualifications _____
fees/notes _____
_____
_____

2. Name of Company _____ point of contact _____
address _____ phone _____
qualifications _____
fees/notes _____
_____
_____

3. Name of Company _____ point of contact _____
address _____ phone _____
qualifications _____
fees/notes _____
_____
_____

CD & I — GM COMPANIES cont.

4. Name of Company _____ point of contact _____
address _____ phone _____
qualifications _____
fees/notes _____
_____
_____

5. Name of Company _____ point of contact _____
address _____ phone _____
qualifications _____
fees/notes _____
_____
_____

6. Name of Company _____ point of contact _____
address _____ phone _____
qualifications _____
fees/notes _____
_____
_____

7. Name of Company _____ point of contact _____
address _____ phone _____
qualifications _____
fees/notes _____
_____
_____

8. Name of Company _____ point of contact _____
address _____ phone _____
qualifications _____
fees/notes _____
_____
_____

9. Name of Company _____ point of contact _____
address _____ phone _____
qualifications _____
fees/notes _____
_____
_____

10. Name of Company _____ point of contact_____

address _____ phone _____

qualifications _____

fees/notes _____

_____

_____

11. Name of Company _____ point of contact_____

address _____ phone _____

qualifications _____

fees/notes _____

_____

_____

12. Name of Company _____ point of contact_____

address _____ phone _____

qualifications _____

fees/notes _____

_____

_____

13. Name of Company _____ point of contact_____

address _____ phone _____

qualifications _____

fees/notes _____

_____

_____

14. Name of Company _____ point of contact_____

address _____ phone _____

qualifications _____

fees/notes _____

_____

_____

15. Name of Company _____ point of contact_____

address _____ phone _____

qualifications _____

fees/notes _____

_____

_____

CD & I — GM COMPANIES CONT.

16. Name of Company _____ point of contact_____

address _____ phone _____

qualifications_____

fees/notes _____

_____

_____

17. Name of Company _____ point of contact_____

address _____ phone _____

qualifications_____

fees/notes _____

_____

_____

18. Name of Company _____ point of contact_____

address _____ phone _____

qualifications_____

fees/notes _____

_____

_____

19. Name of Company _____ point of contact_____

address _____ phone _____

qualifications_____

fees/notes _____

_____

_____

20. Name of Company _____ point of contact_____

address _____ phone _____

qualifications_____

fees/notes _____

_____

_____

21. Name of Company _____ point of contact_____

address _____ phone _____

qualifications_____

fees/notes _____

_____

_____

22. Name of Company _____ point of contact_____

address _____ phone _____

qualifications_____

fees/notes _____

_____

_____

23. Name of Company _____ point of contact_____

address _____ phone _____

qualifications_____

fees/notes _____

_____

_____

24. Name of Company _____ point of contact_____

address _____ phone _____

qualifications_____

fees/notes _____

_____

_____

25. Name of Company _____ point of contact_____

address _____ phone _____

qualifications_____

fees/notes _____

_____

_____

26. Name of Company _____ point of contact_____

address _____ phone _____

qualifications_____

fees/notes _____

_____

_____

27. Name of Company _____ point of contact_____

address _____ phone _____

qualifications_____

fees/notes _____

_____

_____

MARKETING

TIME FOR THOUGHT

1. Describe how your idea evolved.

_____

_____

_____

_____

_____

_____

_____

2. How did the prototype evolve?

_____

_____

_____

_____

_____

_____

_____

3. How do your statistics show that the consumer will purchase your idea?

_____

_____

_____

_____

_____

_____

_____

4. What improvements are needed to better satisfy the consumer needs?

_____

_____

_____

_____

_____

_____

_____

_____

NOTES

NOTES

NOTES

MANUFACTURING

This section on manufacturing is designed to provide an understanding of the requirements associated with the processes of mass producing a product.

By now your prototype has continued its refinement and has evolved into more of a final functioning product to the best of your capabilities. Now it is necessary to determine all the requirements for producing adequate quantities to meet the demand of the market.

> ☛ *RULE # 8 - BE AWARE.*

Take notice of your surroundings and your internal language. Make sure your surroundings and internal language are providing you positive feedback and encouragement. As an inventor, be your <u>best</u> critic.

The Idea LogBook is designed to provide information on developing an idea into a product and therefore will not go into all the intricacies of manufacturing. This section will provide you a wealth of information on the basic requirements and numerous facets of manufacturing; however, the complexity of it alone requires its own separate *LogBook*.

The Manufacturing Process: Let's Take a Look

The very essence of manufacturing is to take and transform or change an entity, by adding additional material, energy and time. For instance, sand is a raw material that can be transformed into glass. Electronic equipment requires numerous components, which are also considered "raw materials" and may be utilized in numerous other products.

Each example requires material, time and energy to transform into and/or add a value for a further use. It also requires change of an existing product into a completed end product, ready for the consumer. This process of changing or transforming is called *production*. Some of the costs include purchasing materials, inventory control, scheduling, routing, loading, warehousing, quality control, and distribution. The costs also include material, labor, machinery, tools, molds, packaging and even overhead such as taxes, building rent and all its associated costs.

Depending on the complexity of your product, each of these above-stated categories are to be included in determining the manufacturing costs. Each step in the manufacturing process needs to be measured and addressed to continually reduce costs and assure a more competitive market price position. There are some industries that have complex processes and vast intricacies making manufacturing a challenging, exciting and rewarding career.

In most cases, the process in which you developed your few prototypes is inadequate for producing the number of products needed to sustain any type of steady sales. The costs for producing your prototypes greatly exceed the costs required to return needed profits.

As you have proceeded through this journey, you should have been recording the expenses you incurred for parts and material, as well as the number of hours you worked, developing your prototype(s). This information is essential. A tally of all your expenses, including labor, will provide a very

conservative cost that you incurred in the development of your prototype. However, unless you are able to determine dental, medical and other insurance or retirement plans, you will not have an actual "complete" market cost. Just make sure to record the figures you are able to get at this time.

If you look at your prototype costs once you've figured in all your time and materials, the price of selling the product at these costs plus profit would be very undesirable in the eyes of the consumer. It is normal for prototypes to cost four times the actual manufacturing costs, if not more.

Proper manufacturing processes are designed to reduce the costs and time associated in producing a product, thus making the price more attractive to the consumer. As "location, location, location" is to real estate, "reduce costs, reduce costs, reduce costs" are to manufacturing.

If your objective is to begin manufacturing your product in the garage or spare room (or wherever), it is imperative to strive for continuous improvements in the reduction of costs. Do this by reviewing and implementing various principles, increasing simplicity and standardizing every possible component; that is, of course, every component not in the area of your invention.

The more you can make your product with existing standard-type components, the better chance you will have of reducing its cost to assure a more competitive product. It is very important to review all areas of manufacturing (for that matter you should review *all functions*) including procurement, fabrication, labor, assembly, engineering, and distribution to see how costs can be reduced and sufficient profits can be obtained.

Steps to Improve Your Manufacturing Process

There are numerous avenues for implementing and improving manufacturing costs and improving processes for almost every type and size of industry. One of the key factors that will help decrease costs during the production processes is to reduce the number of steps and handling of the product. The more times a product is handled, the more it increases the cost. Coordinate the production so that all the tasks are being conducted and the units come together at the same time and you are not waiting on any section for its completion.

If at all possible, during each process, conduct one task on a unit while completing or preparing the next unit. For example, in testing electronic equipment, the process may involve a 10 to 15 second delay to cycle. Use this time to make sure the next unit is prepared for the test. Eliminate any idleness.

If you require knowledge in relation to your specific industry, read books and contact and interview individuals who have firsthand and direct involvement in your product's related industry. The most effective method to do this is to contact several small to mid-sized local manufacturers in your product's related industry and educate yourself with some of their processes by speaking with the person who is responsible for the overall production. This could be anyone from a shop manager all the way up to the president, depending on the size of the company. As long as you will not be directly competing against them with your product and are very cordial and sincere, you should be able to easily meet with most individuals at these companies.

Most presidents and owners are proud of their organization and would be delighted to be interviewed and show off their companies. The knowledge you gain could provide better manufacturing methods and skills for your product that could make or break its success.

Another method to assure your manufacturing costs are in line (or even better) with manufacturers in your product's industry, is to contact several manufacturers who are open to produce other products, and have them propose to you what they will charge to produce your product. They will be able to break the costs down on a per-item basis. Granted, they will have profit included in their proposals, but it will provide important information to assure your manufacturing costs are in line. Furthermore, if you later choose to have the manufacturing of your product conducted by a different company, you now have made viable contacts for an alternate source. Having the manufacturing completed by another company (outsourcing) is a very practical method to mass produce products. It saves you all the immediate costs associated with creating a manufacturing plant until the time when enough sales warrant it.

It also eliminates the possible engineering drawings most products may require because it lets the manufacturer determine for itself what is needed to produce your product, and the costs for these should be included in the price per item. Your prototype or detailed drawings should provide enough guidelines for the manufacturer.

Do I Need to Create Those Complicated Engineering Drawings Myself?

If the complexity of your product requires engineering drawings to be rendered and you lack the skills or equipment to do them yourself, here are a few suggestions.

The first suggestion is to (again) contact small to midsize companies in your related industry and ask if they have the ability to (and would) produce engineering drawings for you. They will, of course, charge for their services depending on the complexity of your product. Therefore, it is in your best interest to shop several companies to assure you are paying a fair market fee.

The second suggestion would be to contact these same sized companies and barter services with them or see if they would be open to receiving a percentage of the product. Of course this will require the creation of a contract about which it would be advisable to consult your attorney.

Either way, these companies have computer-aided design and manufacturing machines (CAD/CAM) that will be able to produce precise drawings to your specifications. These drawings will help you coordinate your concept with what actually can be achieved through the manufacturing process. Sometimes you may have to make a few adjustments on your design to fit into the existing manner in which products have to be machined to achieve maximum returns.

Purchasing Supplies and Other Manufacturing Tasks

Purchasing supplies and components should only be completed when necessary so as not to have too much inventory sitting around. Of course you do not want to run out of supplies and have to stop production. You should be able to determine the exact number of needed supplies by the number of units you expect to produce and the time it will take to complete. Coordinate all purchases with production so that any reorder shipments arrive just in time to be placed in the process.

One of the most important internal tasks you can do is to keep the facility clean and organized. Just the fact that doing so could prevent injuries from employee accidents should be enough motivation; however, it could affect the money you have tied up in inventory. By keeping everything clean and organized (shelves work beautifully), you will be able to have a consistent place for everything and easily locate all required inventory when it is needed for production.

Finally, always strive to improve quality. Quality can make or break a company. It is hard to come back from a negative quality reputation.

The following pages are to be utilized to contact manufacturers should you require the services of these companies.

MANUFACTURING COMPANIES

1. Company Name: .... Date: ....

   Address: ....

   Contact: .... Phone: ....

   Your requirements: ....

   Their recommendations: ....

   ....

   ....

2. Company Name: .... Date: ....

   Address: ....

   Contact: .... Phone: ....

   Your requirements: ....

   Their recommendations: ....

   ....

   ....

3. Company Name: .... Date: ....

   Address: ....

   Contact: .... Phone: ....

   Your requirements: ....

   Their recommendations: ....

   ....

   ....

4. Company Name: .... Date: ....

   Address: ....

   Contact: .... Phone: ....

   Your requirements: ....

   Their recommendations: ....

   ....

   ....

5.   Company Name: Date:

    Address:

    Contact: Phone:

    Your requirements:

    Their recommendations:

6.   Company Name: Date:

    Address:

    Contact: Phone:

    Your requirements:

    Their recommendations:

7.   Company Name: Date:

    Address:

    Contact: Phone:

    Your requirements:

    Their recommendations:

8.   Company Name: Date:

    Address:

    Contact: Phone:

    Your requirements:

    Their recommendations:

Manufacturing Companies Cont.

9. Company Name: .... Date: ....

   Address: ....

   Contact: .... Phone: ....

   Your requirements: ....

   Their recommendations: ....

   ....

   ....

   ....

10. Company Name: .... Date: ....

   Address: ....

   Contact: .... Phone: ....

   Your requirements: ....

   Their recommendations: ....

   ....

   ....

   ....

11. Company Name: .... Date: ....

   Address: ....

   Contact: .... Phone: ....

   Your requirements: ....

   Their recommendations: ....

   ....

   ....

   ....

12. Company Name: .... Date: ....

   Address: ....

   Contact: .... Phone: ....

   Your requirements: ....

   Their recommendations: ....

   ....

   ....

PACKAGING COMPANIES

Once you almost have the manufacturing process worked out, it is time to determine how you are going to package and display your product.

Since you know how the competitors are packaged, you should have an idea as to how your product can be packaged. Be certain to determine the type of box, wrapping, and how it will be protected in a box for shipping so that it will not be damaged.

The best and easiest solution to determine the most suitable packaging for the product is to contact local packaging companies, located in the phone directory, or call one of your now acquainted associates and ask them who they use. Once you have recommendations, you can call these companies yourself and get quotes. Packaging has to be included in your per unit cost.

If needed, let the packaging companies inform you of their suggestions on the best method to package your product. These companies are very competitive and their price will be too. Again, be sure to compare so you get the most for your money and have them complete a confidential agreement.

The following is to be utilized to coordinate and determine the best packaging company.

PACKAGING COMPANIES

1. Company Name: .... Date: ....

   Address: ....

   Contact: .... Phone: ....

   Their recommendations: ....

   ....

   ....

   Type of Material: .... Quantity per box: .... Display: ....

2. Company Name: .... Date: ....

   Address: ....

   Contact: .... Phone: ....

   Their recommendations: ....

   ....

   ....

   Type of Material: .... Quantity per box: .... Display: ....

P ACKAGING C OMPANIES CONT.

3. Company Name: .... Date: ....

    Address: ....

    Contact: .... Phone: ....

    Their recommendations: ....

    ....

    ....

Type of Material: .... Quantity per box: .... Display: ....

4. Company Name: .... Date: ....

    Address: ....

    Contact: .... Phone: ....

    Their recommendations: ....

    ....

    ....

Type of Material: .... Quantity per box: .... Display: ....

5. Company Name: .... Date: ....

    Address: ....

    Contact: .... Phone: ....

    Their recommendations: ....

    ....

    ....

Type of Material: .... Quantity per box: .... Display: ....

6. Company Name: .... Date: ....

    Address: ....

    Contact: .... Phone: ....

    Their recommendations: ....

    ....

    ....

Type of Material: .... Quantity per box: .... Display: ....

TIME FOR THOUGHT

1. What are your total costs for materials?

....

....

....

2. What are your total labor hours per prototype?

....

....

....

3. What is your cost per prototype?

A. Costs for material and parts, etc. + Labor
   (*number of hours worked* x *minimum wage*) ÷ the number of prototypes (1) =

B. If you developed more than 1 prototype with all your material and components, then take the number you calculated in step "A" and divide by the number of prototypes that you did develop.

C. Now add the cost of the packaging and that will equal an <u>approximation</u> of your per unit costs.

   A.

   B.

   C.

   _____

TOTAL

4. How could you improve the manufacturing process to reduce costs?

....

....

....

....

....

5. What manufacturing process would be best for your product?

....

....

....

....

TIME FOR THOUGHT CONT.

6. What other methods could you utilize to reduce costs?

7. What other manufacturing sources are available to you?

8. What are the advantages of outsourcing in relation to your skills and product?

9. Describe how your product will be packaged for shipping. Number of units per box?

10. Describe how your product will be packaged for display. Number of units per box?

11. What is the total cost for packaging?

NOTES

MANUFACTURING

NOTES

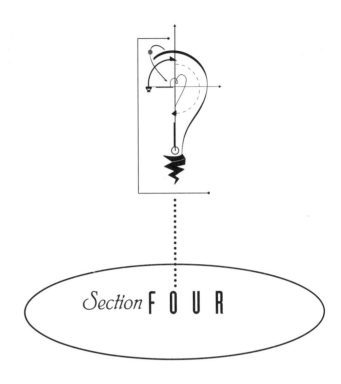

Section **F O U R**

In this section we will look at
the possibilities available to you
for bringing your new concept to
mass production.

NOW WHAT?

Okay, now you have conducted research into the market, the industry, determined the existence of a market, developed a working prototype, and protected the idea. You now understand the costs associated to manufacture the product and have determined a very favorable overall potential.

Now What? Where do you go from here?

Here is where it begins to get even more fun and exciting.

> ☛ *RULE # 9 - ACCEPTANCE.*

Accept others for the gifts they bring and yourself for the gifts you bring. Everyone is unique and special — including you! Understanding who you are and knowing your purpose will provide strength and insight into why you are special. This will help you see the strength in yourself and others.

There are two basic directions you could take:
1. License the idea to a marketing or manufacturing firm.
2. Establish your own business then develop and market the product yourself or outsource portions of it. Either of these directions offers other various avenues which have a completely different outcome.

A careful decision is required based on your capabilities, expectations, personality and especially the reasons as to why you are taking up this journey. Return to "Conceptualization" (Chapter 2), and review and update your reasons for beginning this journey.

This *Now What* chapter is designed to assist you in taking a personal inventory of your capabilities, including your strengths and weaknesses.

Turn the page and answer the following questions:

PERSONAL INVENTORY

Complete the following list by placing the letter A, B, C or D by each word that best describes your strengths and personality traits. The letter "A" indicates a strong ability where a "D" indicates less ability. There are no wrong answers!

This is not a complete list or one that will decide the success of your idea, but rather a chance for you to increase your awareness so as to make a well-informed decision.

___ Self-starter

___Communication skills

___Ambition

___Responsible

___Taking risks

___Need to be in control

___Financial resources

___Goal setting

___Task-oriented

___Attitude

___Hard worker

___Listener

___Diplomatic

___Resourcefulness

___Faith in self

___Visionary

___Humility

___Fairness

___Creativity

___Management skills

___Business skills

___Organized

___Leadership skills

___Credit rating

___Trust level

___Support of family

___Willingness to work 60 or more hours per week

___Willingness to work every position and then some

___Persistent

___Independent

___Strong-willed

___Disciplined

___Faith in others

___Open minded

___Sincere

___Respectful

___Enjoys people

___Marketing skills

___Desire to be own boss

___Accept loss

___High energy

___Make decisions quickly

___Innovative

___Time management

___Cheerful

___Competitiveness

___Confident

___Sociable

___Service oriented

___Willingness to learn

___Honesty

This list is only to give you a basic understanding of some of the requirements necessary to become an entrepreneur. It is to aid you in understanding your strengths on which to capitalize, areas where improvements may be needed, or where outside assistance should be obtained.

The more spaces with A's or B's indicates a strong potential in your personality becoming an entrepreneur and creating your own business, which demands a much stronger determination and drive.

If you have more C's and D's, it is a strong indication that you need more education and knowledge in creating and operating a business or that your idea may be better off in the hands of much more qualified individuals and companies.

However, the truth of it boils down to, where do *you* choose to go? Determination and ambition are very powerful allies that could be more than enough to make you extremely successful, no matter what any test indicates or anyone says.

Creating your own business *does* require more time and effort than licensing and if your overall expectation is to create a business, go to "Creating a Business" (Chapter 11) and begin. Also, go to Chapter 13, "The Final Suggestions" as it provides a clear picture of the options that are available to choose.

If your overall expectation is to turn your idea over to more experienced individuals and collect a royalty, then go to "Licensing" (Chapter 10) and begin. Licensing your idea allows you to focus all your attention on product development and not dividing it to also create a business.

To better assist you in your decision, take a few minutes and read the next two chapters, as they will explain in more detail about the tasks involved.

Choosing to either license or create a business is a personal decision based on your goals and expectations of your product. Either choice is a sound decision, versus not making any decision and not even developing your idea.

If, on the other hand, you choose to take the "out door" and exit from developing your idea (hopefully only because it already exists) then go to "Conclusion" (Chapter 14). The only wrong decision would be to do nothing. Don't take your idea to the grave with you. Let everyone share from it.

T IME FOR T HOUGHT

1. What were your original reasons for developing this idea?

....

....

....

....

2. Have any of these reasons changed? Why? If so, explain how and why.

....

....

....

....

3. What were your overall results from the test?

....

....

....

....

4. What are the areas of your strengths?

....

....

....

....

5. What are the areas of your weaknesses?

....

....

....

....

6. What actions can you take to eliminate your weaknesses or turn them into an asset?

....

....

NOTES

LICENSING

How do you like the thought of creating an idea and then selling it to someone who mass produces it while you collect royalties? Can you see yourself grinning as you sit on the beach sipping tropical drinks on your own private island? This is almost everyone's fantasy. This is called licensing.

Though this is not impossible, by now from all your time, efforts and education, you realize that in order for it to happen, you must take the right action.

We currently live in a throw-away society. Once we purchase something, in a very short period of time an improved version is introduced and we immediately begin convincing ourselves of the reasons why we must have it. Computers are a great example of this. At the current pace they are improving, it appears as if they are outdated by new and improved models before you can even get your newly-purchased one home and out of the box.

Products have a certain life span called the "product life cycle." This cycle consists of its introduction, growth, maturity and decline. Each industry's product life cycle varies. From the time a new product is introduced, its end is in sight for the next better edition.

This is a throw-away society with a short-lived product life cycle. These two facts provide two reasons why you should continue to remain open for receiving new ideas.

Companies are well aware of the throw-away market and product life cycles and in order for them to maintain a strong position in the marketplace, it is imperative that they continue to introduce new products along with improved versions of their existing ones.

New products are the life-blood of the market.

This section will provide you with the skills necessary to contact individuals within companies and determine if they have an interest in your idea. It will also continue to educate you as to the best possible direction to pursue with your idea.

Perhaps You Can Be That Inventor Sitting on a Beach Somewhere Collecting Royalties.

The greatest challenge for inventors is educating themselves about the processes involved in bringing their idea to the market. Again, remember coming up with an idea is the *easiest* part of inventing. The excitement and challenge lies in getting it to the market.

The best way for inventors to quickly get their products to the market is to leverage the assistance of others who have the experience, knowledge, market position and finances. They do this through licensing their ideas. You can do this too.

☛ RULE # 10 - BE HUMBLE, NOT RECKLESS AND ARROGANT.

You can get everything you choose without being rude or having to hurt anyone.

This Section Is Based on Timing and Synergy.

Timing -   The company, market and economy is ready for your idea.

Synergy -   Two or more individuals working together in harmony to achieve a common, equal, fair and desired goal.

There are three key areas that destroy synergy. They are greed, attitude and the inventors' emotional attachment to their ideas—being afraid to let it go.

There have been many incidents where a company was willing to play ball and run with the inventor's invention, only to have been negatively blasted by the inventor's ego of greed, followed by his poor or hostile attitude and then his not wanting to let the product out of his control. You will be giving up some (or all) control and ownership of the idea, *but so what?* In order to see it placed on the shelves of the store, to make society a better place and earn royalties, isn't it worth giving up some (or all) control?

Fact for Thought:
Giving up control and letting something go will allow you to receive even more in return.

Business people will only put up with so much before they throw in the towel and move to a different invention or product that fits into THEIR synergistic environment. Go back to "Conceptualization" (Chapter 2) and again read the reason(s) why you are on this journey. Write any changes that have come about since your original thoughts and any additional reasons that have surfaced as to why you are developing this product.

Companies are focusing on how much money this product will produce for them and are in business to make a profit. Your own reasons for creating this product need to be clear. To increase the chance of your product's success, it is imperative that your attitude, emotional attachment and greed *not* play a role by destroying the SYNERGY between you and the company that will manufacture and market your product.

Who Will Want My Idea?

There are three places to submit your idea for someone to review and then determine if it fits into their product plan. It must also provide them with the desired and expected profits to make the venture worth their while. These three places are:

1. Manufacturers
2. Sales or Marketing Companies
3. Existing Wholesale and Retail Businesses

1. Manufacturers

There are manufacturers who actually make, sell (in-house contracted sales force) and distribute their own products, and are interested in purchasing the ideas of others to manufacture. In other words, they have the resources, tooling, established networks and expertise to produce and market products.

You can be certain that many manufacturing companies are looking to grow and expand their producing capability through the addition of more products. It is in their best interest to at least review potential products. The key to locating the best manufacturing companies is not to contact the large companies because they usually have an in-house research and development department (R&D) creating their own new products, and besides, the bureaucracy will only leave you frustrated.

The key is to contact small to midsize companies with a strong marketing structure who cannot afford, or do not want, a full-time R&D department. The strength these companies have is their potential to remain extremely competitive because of their lower costs of doing business. They do not need an R&D department if they take advantage of the pool of ideas that are available from inventors like yourself. They are also more likely to be able to move quickly since they are not dealing with layers of bureaucracy that must get involved in the decision-making process.

2. Sales & Marketing Companies

The second method (and by no means are these listed in order of importance) for licensing your product is through pre-existing sales and marketing companies. Sales and marketing are the most important factors in getting a product sold. These companies excel in this area and utilize manufacturers as subcontractors for producing their products.

It is extremely important that you evaluate marketing firms with a fine-tooth comb. There are some good ones, but be aware that if they work off your ego or your emotional attachment to the idea, they can give unlimited promises and leave you with nothing.

The most reputable companies will NOT require up front, advance money but rather a percentage of the profits. By you paying up front, they have nothing to lose, no risk, and the incentive to do anything still lies in the hand of the inventor—you. To assure you have the best marketing firm to meet your requirements, ask for a track record, get knowledge of all fees; understand completely what you get for your money, review products they have developed (which should be proudly on display). Contact the Better Business Bureau and state agencies for any possible complaints filed.

Careful thought and planning are required to license your product to manufacturers and marketing companies. To be sure that they are capable of completing the developing and marketing of your product (and others), do your homework so that everyone makes a profit. Also, it is imperative that you contact companies who are in the same market as your idea. For example, do not contact companies who manufacture electronic equipment if your product is a card game.

Take heart. A tremendous learning curve on your part will be undertaken as you learn and determine which companies will or will not accept outside submissions of new ideas. Many companies feel the need to keep all newly-developed ideas in-house because they feel they are more qualified to come up with better ideas (ego) or they are not knowledgeable on the process of receiving outside submissions or afraid of taking the risk (closed-minded). There are even some who feel that all royalties are to be kept in-house (greed). So be prepared for rejections, but don't take them personally. Don't let your feelings think it is you they are rejecting. Keep moving until you reach the ones who will play. They are out there. By now you should be able to easily locate these companies (library, referral, etc.).

3. Existing Wholesale and Retail Businesses

If your product is complete, developed and already in production, then existing wholesale and retail businesses may be the best direction for your product. These businesses come in all shapes and sizes. You should easily be able to locate several to review your product because they usually are open to receiving outside submission of ideas. This is the same approach as stated in Chapter 7.

The greatest drawback to some of these companies, especially the giants, is they are so inundated with thousands of new product ideas that there is a great chance your idea could get lost in the shuffle. There are some companies with rooms filled with new ideas sent to them just collecting dust.

Let's Discuss Royalties a.k.a. "About That Private Island...."

You are probably wondering what took so long.

Since your idea is to be submitted to someone else who will take it and run with it, there are a few things you need to understand about companies to whom you will be licensing your product. This will explain why it was suggested to focus on profits throughout this process.

First, companies take a huge risk in bringing new products to the market. When they review a product for licensing considerations, they should conduct their own research.

They need to know:

- your product works
- that it's legal,
- that the product fits their product profile
- all costs associated with its development, including material, labor, tooling, overhead, etc., and
- *that it provides a desired return of profits*

To complete product development and then launch a product can take years depending on its complexity. Within that time frame, the market or economy could turn in a new direction and they would be out a substantial sum of money, time and effort.

With your understanding of the risks associated with a company licensing your idea, expect the royalty rule of thumb to be in the range of 3 to 7 percent and sometimes even as high as 10 percent. This percentage will be based on gross or net sales of the product and will not include sales associated with any shipping and handling costs or taxes.

Generally, the higher the royalty rate, the higher the expected gross profit margins and the lower the royalty rate, the lower the expected gross profit margins.

The Proposal

To sell your idea, you will be contacting manufacturers and/or marketing companies. Your first goal is to obtain an invitation to present your product. It is best to begin developing a strategy and proposal for that presentation.

Each company that extends you an invitation to demonstrate your product will most certainly have different requirements, information and methods for reviewing outside submissions. Your best tool for understanding these requirements is to begin developing a desired synergy between you and the company by asking as many questions as you can so you are clear as to how they want the demonstration presented.

There are some basic certainties required in the proposal and; from your questions, you should be able to refine it easily to meet THEIR requirements. Make sure you are clear as to who, what, where, when, how and why.

The following page is a suggested outline to begin preparing the proposal. Add, delete and arrange the topics to fit your product characteristics and the company's requirements. It must be typed; drawings are to be crisp and clean; and it needs to be placed in a conservative cover or binder. Make as many copies as there will be evaluators for each company.

As you read this outline, you should notice that you have actually completed most of the work already. Here is where all your efforts have paved a way to extreme simplicity. Just go to the various sections of the *LogBook* and place all your answers into the proposal. Also, *The Idea LogBook* provides valuable information as to the development process that can validate your outline, and it should be brought to the presentation for review by the company.

Name of Company:

THE PROPOSAL

Number/Names of those in attendance:

Cover - indicate the name of the product, state that you are presenting to (name of company) and that it is being presented by (inventor's name).

Table of Contents:
1. Summary:
   a. name
   b. description
   c. function
   d. benefits
   e. development history

2. Costs:
   a. material
   b. assembly/labor
   c. production

3. Patent filings:

4. Background on self:

5. Competition:

6. Market tools:

7. Miscellaneous information you feel adds value:

8. What else?

9. What else?

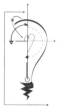

The Contact

You should be very proud of yourself for reaching this stage of developing an idea into a product. You truly are an inventor!

A great deal of the work has been completed and now it's time to prepare the idea so it can complete its next phase as it heads to the market. An analogy of this phase could be described as helping pick the college for your child and then preparing them for the journey of sending them off alone into the world with your blessings and support.

Keep up With Your Research

By now you should have compiled a list of companies to contact and know to return to the library (with that friendly librarian, who you now owe your first-born's name to), its very friendly computer and the various registers for thousands of additional prospective companies to contact. In addition to easily obtaining the required information through these avenues, public companies will be happy to send you, for free, their annual report. Just call them and request it for investment considerations. *And you are considering investing in them.*

Inside the annual report of public companies is information that directly focuses on their organization structure, type and depth of products, how they market their products, their advertisement costs, R&D budgets and expenses, royalties paid out, and even pending lawsuits. Private companies, however, require a bit more effort for obtaining information. No matter how much time and energy you put into researching them, your information will not be as in-depth as with a public company.

The type of information that should be easily accessible about private companies is their type of business, when they organized, the name of owner(s), their credit references, number of plants, sales, employees, product lines, distribution, marketing and advertising. The Better Business Bureau, business section of newspapers and business magazines may also provide some information.

Also, check the computer at the library or even the Internet for news articles or stories on companies in your idea's field. If a company is looking to generate more growth through new products and they want to get these ideas from outside sources, you might have to read between the lines, but it will be stated.

Your list should grow substantially from here, as your purpose is to quickly learn who is open to review outside products to mass produce and market.

Now It's Time to Begin Contacting Companies

As you begin this process, your tongue may be very heavy ... so much that the words coming out of your mouth may seem unclear. The phone may feel like it weighs a thousand pounds. The best and only suggestion to easily overcome this is to focus on your product being in the market. Keep pressing on. Begin with a goal of spending a certain amount of time each day making a certain number of calls.

The more calls you make, the more comfortable you will become and the conversations will become smooth and the phone will feel like a feather. Guaranteed.

Understand that more companies will reject the idea of reviewing outside inventions than ones that will be open to reviewing them. The reasons are many, as was discussed earlier, but the more "No's" you get, the closer you will be to the one(s) that says *YES!*

Again, do not take "No's" personally. The company who does say no is not rejecting you or your idea. They will just be rejecting the concept of receiving outside submissions for new products. Do not let your negative emotions interfere and stop you after a few calls because your feelings got bruised from a couple of "No's."

Be Persistent!!! It may take a hundred calls, but there are *YESSES* out there from companies who will be happy to review your product and, quite possibly, provide you all the success you've dreamed of. Just keep reaching for it!

What Do You Say? ... and to Whom?

Below is a script explaining a basic understanding of what to say and expect when contacting companies. Though every conversation will be completely different, it is designed to quickly get to the point and then get the answers you need. Modify it into your personality and product as needed.

No matter how the conversations turn out, *be very cordial and polite* to everyone you speak with because some of the smaller companies have employees wearing numerous hats and the person answering the phone could be the president's assistance or maybe even the president!

Secretaries are extremely influential individuals within a company's organization. Besides their regular duties, they know valuable information on the company and provide their boss with much support, which can include arranging schedules, prioritizing messages, and expressing opinions (feelings) about *somethings and someones.* Remember that they spend a great deal of time with their boss, to whom they are very loyal.

Get to know receptionists, secretaries and any personnel of any company with whom you come in contact (like you did with the friendly librarian). They can make or break the success of your invention at a company.

This does NOT mean you go out and buy a dozen roses or box of chocolates for them. It means be polite and friendly. You'd be surprised how far a pleasant disposition will get you into the President's office.

Your conversation most likely will not flow this smoothly; the script should be customized to fit your personality. Just remember, you are first asking a company if they are open to review a product that fits their product line; and secondly, for an invitation to present your product.

S A M P L E   S C R I P T

Company Operator - *"Hello, XYZ Company"*

You - *"Hello, my name is _____, with whom may I speak about submitting a new product to your company?"*

Operator - *"Where you calling from? / What company are you with?"*

You - *"Name the city you live in. / I am not with any company, I am from (city)"*

Operator - *"Just a minute, let me connect you."*

You - *"Thank you."*

Bob - *Bob here.*

You - *Good morning Bob, I was directed to you as the person responsible for receiving outside submissions of new product ideas, is this true?*

continued on next page... The excitement mounts...

SCRIPT FOR CONTACT

Bob -   *Yes.*

You -   *Great. My name is _____, is your company currently*

            accepting any new ideas from outside the company?

Bob -   *Yes.*

You -   *My new product idea is along the line of _____ (describe its general charac-*

            teristic, i.e. exercise equipment, board game, etc.), would you be open to review this type

            of product to possibly add to your product line?

Bob -   *Yes.*

You -   *Is your company open to a confidentiality agreement?*

Bob -   *Yes. Also, we prefer an item patented or pending, but will entertain others depending*

            on the product.

You -   *What is your process for preparing for the presentation?*

Bob -   *Well, you just come 15 minutes prior to the presentation; my secretary*

            handles the paperwork and coordinates the process.

You -   *I have a proposal and prototype. Is there anything else your company requires?*

Bob -   *No.*

You -   *What do you require to be in the proposal?*

Bob -   *Summary, history, costs, background, patent information, the usual.*

You -   *How often are presentations conducted?*

Bob -   *Every 4th Tuesday of the month.*

You -   *How do I place myself on the next available opening?*

Bob -   *What is your name?*

You -   *(Say and then spell your name).*

Bob -   *What type of product do you have?*

You -   *(Again, describe its major characteristic) Exercise equipment, etc.*

Bob -   *Okay, done. You are set up for the 4th Tuesday of next month.*

You -   *Great. Thank you for your time.*

Once you have successfully made an appointment with a company, do not stop making calls. Having only one prospective company is like having one tire on your bike—you could make it work, but why not ensure a smooth ride? There are just too many unforeseen variables that can arise to stop after one appointment.

Continue contacting companies until you have several companies with whom you will be meeting to present your idea. This puts you in a stronger position to ensure you are getting a fair deal and increase the chance of your idea getting to the market. It also eliminates some pressure off you to make that *one and only deal* happen. Furthermore, it is in your best interest to have more than one company to educate yourself on what the industry will bear. For instance, maybe the royalties in your product's industry are averaging closer to ten percent. Wouldn't that be important to know in your negotiations?

This does not mean that you should play one company against the other to squeeze out everything and anything you can. Companies will be able to detect this easily which may result in you having all of the companies backing away because of *the greedy inventor*. Your success is based on the success of the product in the market, as you will be earning a percentage of its sales. Having a company that fits your personality and goals is your highest priority, even if they pay 1 or 2 percent less. Peace of mind is worth it.

Prior to making any presentations, it is very important to gather as much information and knowledge as possible about that company prior to the meeting. The more you know about the company, the more you will understand its operation and business objectives. This also shows those with whom you meet that you've already invested in them, taking the time to learn about their needs. Return to "Marketing Evaluation" (Chapter 5) if needed, and conduct thorough research about this company as you are able.

And if I Get the Deal?

If after your first presentation you and the company happily agree on a deal, contact the other companies with whom you had arranged subsequent meetings, thank them for their time and opportunity they were allowing, and let them know that you have accepted an offer already. You never know, you just might need them for your next idea.

If the other companies ask for details about the product and the company that picked up your product, be polite and cordial but say nothing. Remember that they are now the competition. Your alliance is elsewhere, and there is no need to take a chance on ruining or sabotaging your new existing deal and relationship.

The following pages are additional forms to utilize to maintain organized lists of the companies you have contacted. At this junction, the only information required is just company name, address, phone and point of contact. If you prefer, while at the library, just copy the pages from any of the numerous books or print the lists generated off the computer. If you choose to utilize these pages and you will require more, just copy the last page prior to using it.

PROSPECTIVE LICENSING COMPANIES

1. Company Name
Point of contact & Title
Address
Telephone
Notes

2. Company Name
Point of contact & Title
Address
Telephone
Notes

3. Company Name
Point of contact & Title
Address
Telephone
Notes

4. Company Name
Point of contact & Title
Address
Telephone
Notes

5. Company Name
Point of contact & Title
Address
Telephone
Notes

6. Company Name
Point of contact & Title
Address
Telephone
Notes

7. Company Name
Point of contact & Title
Address
Telephone
Notes

8. Company Name
Point of contact & Title
Address
Telephone
Notes

9. Company Name
Point of contact & Title
Address
Telephone
Notes

10. Company Name
Point of contact & Title
Address
Telephone
Notes

11. Company Name
Point of contact & Title
Address
Telephone
Notes

12. Company Name
Point of contact & Title
Address
Telephone
Notes

The Prototype

The prototype is a very important piece to the puzzle when conducting a presentation, as it will increase your product's chances of success. All measures should be taken to ensure its readiness and that it performs exactly as you claim.

The advantage of having a prototype is that it proves your idea can be produced as a tangible product and allows the examiner to incorporate all their senses (that can be) into its review. The ability for someone to see, hear, touch, smell and taste reinforces the prototype's validity.

To illustrate this point, go to your local mall and observe a young child and their parents as they shop. Notice how the parent will constantly instruct the child "don't touch that" or "leave it alone." In essence, the child is just trying to satisfy their curiosity and grow and expand their horizons through the natural use of their senses ... in this case touch or feel. During the examination process you will observe adults acting in this same manner as they evaluate your prototype. The proposal will be examined thoroughly for its validity, but the prototype will be ever so carefully scrutinized through sight, touch, sound, smell and even taste.

Your prototype will be picked up, handled, manhandled, edges pulled, corners jabbed, banged and dropped, etc., and if there are any weaknesses or flaws, there is a good chance of them surfacing.
If you can, bring at least two prototypes to the demonstration. If your product has multiple units, bring at least one of everything. For instance, if it has a set of 4 different robots, bring one of each.

If, for some reason, you were not able to develop a final prototype made exactly to the end product specifications, develop large drawings on poster board (see a graphic artist). These drawings must be of high quality, clear, crisp, simple to read, and use the same colors you would have if you had the prototype.

Depending on your type of product, you may also consider creating drawings to add to the presentation. While your prototype is being passed around and examined, the drawings may provide you precise sections to point at, better assisting in answering questions.

Instructions   - If your product requires it, make certain they are included.

Packaging   - If your product requires some type of packaging (box or carton, etc.) it is in your best interest to have it packaged. If it requires special packaging, it absolutely has to be included. Some companies may not be concerned about packaging as it will allow them to design something more appropriate to their marketing strategy and company profile.

The Presentation

The presentation can be the most rewarding or nervous time for an inventor. With the work that you have completed and by following these basic principles, it will be mostly rewarding.

You now have the right number of copies of the proposal, a prototype or drawings, an understanding of the requirements, guidelines, and methods the company expects and, therefore, the presentation can easily be planned.

Now it is time to prepare one more thing a company looks at when considering licensing a product - *you, the inventor.*

It is very important for a company to feel that the inventor is going to *play ball* with them rather than be a thorn in their side. There have been many deals killed because of the inventor's terrible attitude, disposition and tone of voice.

• Be honest, up front, polite, pleasant, friendly, well-mannered, cleanly dressed, well-groomed and above all, SMILE. It is best to be a little conservative so not to take the focus away from your product. Leave the streamers, balloons, and marching bands for parades. Let your product be the center of attention.

• Get to the meeting location at least 15 minutes early which assures your place (do not let being late ruin your success) and gives you time to observe and become familiar with the surroundings.

• Conducting a presentation is like giving a speech. Many people find it too difficult and scary. Adding the fact that you are emotionally (hopefully not too) attached to your product can increase your nervousness and decrease the success of your product. A remedy to this is to have someone else (perhaps your patent attorney or agent) conduct the presentation while you go along as the "side-kick" for support. Many companies do not look down on this practice because they do not want to deal with the inventor and their emotionally-opinionated outlook.

Someone else conducting the presentation may be able to remain calm and focus on the end result of creating a true synergy and *win-win* situation, especially if the presentation takes a sour turn. Remaining calm and clear-headed will increase the chance of the situation being brilliantly turned back into a more positive direction. If you choose to have someone else conduct the presentation, follow their lead and be polite and friendly. Only speak when spoken to. Silence is a sign of strength.

• Usually you will go into an office or meeting room and sit down to demonstrate and discuss your product. Depending on the complexity of your product, the meeting could only last 5 minutes. The length of time may not have a bearing on its success.

• As you are handing out the proposals and bringing out the prototypes, simply introduce yourself, thank them for their time, introduce the name of the product, give a brief summary of its market segment, manufacturing and production costs, potential profits and patent stage. Also, if the prototype is not made from the final material, explain why, describe the recommended choice of material and cost of developing. Furthermore, be sure to explain who will buy this type of product, why, and at what cost.

• Another very important factor the company licensing your product will require is the extent of the rights or exclusivity that surrounds what they will actually own of it. They will want to know if they will be purchasing the complete territorial market rights, type of use or performance and even the rights the inventor expects to maintain and rule they will play. Make sure you are clear with these areas.

Upon completion of the presentation, the company may immediately inform you as to their decision on whether they will run with your product. Most likely they will discuss it among themselves in a private session and inform you at a later date.

Be sure to thank them for their time and allowing you this opportunity. Politely leave and follow up with a thank you letter.

After the Meeting

Once the negotiations are complete and a company decides to run with your product, then a licensing agreement will be drawn up for review. Signatures will be collected to assure that everyone understands each detail and that everyone is protected. It is highly recommended that you seek proper legal counsel to review the contract and clearly explain its true meaning. Proper counseling (patent attorney?) can make certain items such as auditing, royalties, development, litigation, sublicensing, and much more will be included in the contract to document everyone's rights and responsibilities.

After everything is said and done, it is very important to remain in contact with the company and continue to develop a positive synergy. This can be accomplished by contacting them to say hello, status how things are developing and offering your expertise or services, if not already part of the deal.

Remember that your income is based on a percentage of the product's success and a positive synergy between two individuals striving in the same direction for the same goals can only add value and increase your chances of success.

Following is a checklist to assist in ensuring you and your presentation are ready for the meeting that could change your financial outlook.

THE PRESENTATION CHECKLIST

1. _____ three appointments (minimum)

2. _____ clear as to company requirements for presentation

3. _____ clarity on presentation

4. _____ prototype completed

5. _____ drawing completed (clean, clear and colorful)

6. _____ proposals completed

7. _____ sufficient proposals copied

8. _____ conservatively dressed and groomed

9. _____ attitude in check, be pleasant, polite and friendly

10. _____ presentation rehearsed

11. _____ company researched

12. _____ synergy attitude in check

13. _____ your expectations

TIME FOR THOUGHT

1.   What are some things you can improve to create better synergy?

   ....
   ....
   ....
   ....
   ....

2.   What have been some of the obstacles preventing you from contacting companies?

   ....
   ....
   ....
   ....
   ....

3.   How can you improve this area?

   ....
   ....
   ....
   ....
   ....
   ....

4.   What are you doing to remain positive and focused?

   ....
   ....
   ....
   ....
   ....
   ....

5.   How can you revise the script to fit your personality?

   ....
   ....
   ....
   ....
   ....
   ....

TIME FOR THOUGHT CONT.

6. How can you remain professional during the presentation?

....

....

....

....

....

7. What are your expectations in the contract?

....

....

....

....

....

8. Will you offer to provide technical or other services to the company? If so, describe.

....

....

....

9. What will you provide the company?

....

....

....

....

....

10. Read book on licensing and negotiations
    Name of book & key points to remember:

....

....

....

....

....

....

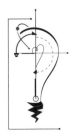

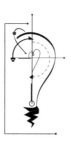

NOTES

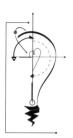

NOTES

CREATING A BUSINESS

Business is the cornerstone of our economic system because it offers everyone the opportunity to organize and operate, for profit, a trade to produce goods or provide services in a competitive environment. Everyone at some time or another has grandiose dreams of creating and running a business and enjoying the reward of being an entrepreneur and the boss.

The main objective of every business is to make a profit. Starting a business is as easy as going to your local city hall and obtaining a business license. However, if you plan on staying in business, your business must generate a steady stream of profits. In other words, it must bring in more money than it spends.

> ☞ *RULE # 11 - REJECT THE CURRENT STATUS QUO.*

Strive for something greater, changing the rules is a good thing. Dare to be different, for it is only through being different that we can get different results.

The beginning and early stages of a business usually are the most challenging in the area of generating sufficient profits to cover the startup and operating costs. A business must be able to endure this period as it takes time to generate sales growth.

There are three types of businesses:

1) sole proprietorship,

2) partnership, and

3) corporation.

In most cases you will be creating a sole proprietorship or partnership business; however, it depends on your needs and characteristics.

Sole Proprietorship

The first type of business, sole proprietorship, is the most common. It is defined as a business owned by one person. It is the easiest to establish. All profits go to the owner, the owner has total authority over the business, it has no special legal restrictions (other than the general criminal and civil laws) and it is taxed as an individual.

The drawback of a sole proprietorship is that the owner may have limited capital available, is liable for the amount invested in the business as well as all other assets, and the length and life of the business is completely dependent on the owner.

Partnership

The next type of business is partnership, which consists of two or more individuals and can be created around anything upon which the two individuals agree. The advantages of forming a partnership are the entity's ability to raise more capital and have additional expertise to support its operation.

There are three types of partnerships: 1) general, 2) limited, and 3) silent. The primary difference between the first two is that the limited partner has limited liability and can only lose their investment in the partnership. A silent partner usually stays away from any day-to-day operations and assists main-

ly with finances and behind-the-scene support. If your decision is to establish a partnership, it is in the best interest of all parties to have an attorney draw up "Articles of Partnership Agreement." This agreement spells out each partner's contribution, authority, profit and loss distributions, and even dispute mediation.

It is recommended that the agreement contain at least the below information but this is not a complete list. Your product and situation will dictate what is best.
1. Date of partnership
2. Name of company
3. Name and address of partners
4. Nature and scope of business
5. Location of all business activities
6. Length of partnership
7. Contributions of each partner
8. Disbursements of profits and losses
9. Salaries
10. Employment hours
11. Authority relationships
12. Access to books and records
13. Arbitration for disputes
14. Terms and method of dissolving partnership and assets

Corporation

The third type of business is a corporation. Corporations have a minimum of three owners (in most states) called shareholders and are considered an *artificial being* in that it may own property, enter into contracts, be liable for debts and sue and be sued. The majority of the products we purchase come from corporations.

A corporation's primary advantage is its ability to accumulate large quantities of money for investment into its assets through the sale of stocks and bonds. The disadvantages are taxes and legal restrictions.

To better assist small businesses, Subchapter S Corporations were added so corporations under certain conditions can be taxed as a sole proprietorship or partnership.

There is also a type of corporation called a Limited Liability Corporation or *LLC*. This type of corporation is very similar to a sole proprietorship and partnership but it has limited liability on behalf of the owner(s) and states that the owner's personal property is protected from any suit, should they occur.

Before choosing the type of business you want to start, get advice and guidance from an accountant or attorney and determine which one is best for your particular business.

As a business owner, circumstances may require you to perform every and all functions of its operations as well as its overall activities of hiring and training employees, purchasing products and services, raising money, accounting (payroll, taxes, receivables, payables, etc.), market and marketing analyzing, manufacturing, distributing and selling. There will be times when you will be responsible for performing each of these tasks on any given day.

Actually Starting Your Business

The Business Plan

The first step in starting a business is to develop a well thought out, comprehensive guide or plan of action for the overall direction of the business. This written document is called a business plan.

The business plan can also be used as a tool should you need to obtain financial assistance from lending institutions.

Below is a suggested outline of a business plan for your consideration. It consists of the minimum requirements for a plan, but be sure to customize it to fit your company and its goals and objectives.

Upon completion, go to your local stationary store and purchase quality paper to print it on and a conservative cover. Many of your answers are already completed from previous sections.

Take your time and carefully complete the plan and when you have finished, walk away from it for a few days and then return to it and give it another edit. And then another....

YOUR BUSINESS PLAN

1. Cover Letter - indicate loan amount expected, length of loan, why the need for a loan and type of collateral. Input your logo (if have one) and use a higher quality paper.

2. Cover Sheet - indicate "Business Plan for Company Name," the name and address of owner(s) and the logo.

3. Executive Summary - answer what you plan to achieve, objectives you expect to reach and how, who will be responsible for these objectives, and capitalization that will be required. This is very brief and completed in less than a page.

4. Table of Contents - this is a reference as to where everything is at in the business plan. Complete this section last and insert it after Executive Summary and begin the numbering sequence with Business Identification.

5. Business Identification - what your business does, its name, address, phone, facsimile (fax) number, tax and business identification numbers, and your attorney's, banker's, accountant's and insurance agent's name, address and phone numbers.

6. Business Purpose - list the goals of the business and how the funds will be utilized and the dollar amount indicated (equipment $2000, inventory buildup $4,000, etc.). Also, depending on your product and business, indicate the amount of employment you will be providing and how the market conditions will improve.

7. Description - describe what the business is, its type (sole proprietorship, etc.), classification, date beginning operation, current status of the industry schedule, plan for managing, number of personnel, list of inventory, list of vendors you will be selling to, list of suppliers and explain the terms.

8. Market Research - needs and wants of customers, market segments, promotional strategy, market trends (growth or decline), sales research (where and how you will be generating leads), and pricing strategies.

9. Competition - names of nearest competitors, estimation of their market share, explain how you are different and better and will be able to earn some of that market.

10. Management - name of each principal individual, title, job description and salaries including benefits, number of part time employees and their salaries, and succession of persons who will run business in the event you become ill. One of the key reasons businesses fail is poor management. Don't let yours fall into this category.

11. Location - list address, why it was chosen, location of nearest competitor(s), type of neighborhood, zoning restrictions, terms of lease, taxes, and floor plan.

12. Marketing Strategy - advertising, sales, promotions, public relations, product costs, desired profits and strategy, who are customers, and products and services offered. Focus on satisfying the needs and wants of the consumers.

13. Financial Information - list capital and assets, depreciable assets, balance sheet, break even analysis, income projections, cost flow projections and operating statement.

14. Record Keeping - explain how you will keep track of day-to-day sales as well as weekly, quarterly and annually, cash sales, credit card sales, customer records, sales promotion and results, personnel costs, expenses, equipment acquisitions, and lease and loan schedules.

The Idea LogBook is designed to assist you in developing an idea of a product and therefore will only *briefly* explain some of the basic requirements to create a business. It would be in your best interest to return to your local library and pick up a book (or two or ten) on starting a business for a more in depth education.

Where to Get More Info

There are numerous resources at your disposal that will provide you with a tremendous amount of assistance with your business for free. A good place to begin is with the Small Business Administration (SBA) which provides financial, technical and management assistance to help start, run and grow a business. Contact the SBA at 1-800-827-5722 or www.sba.gov for the office in your area and to obtain further information about their services.

The SBA sponsors a group called the Service Corp of Retired Executives (SCORE) who are active retired men and woman who provide free management counseling to small business owners and those considering starting their own business. There are over 80,000 SCORE offices throughout the United States which can be located and contacted through the SBA.

Another great resource is trade associations. The greatest advantage of trade associations is that they have a tremendous amount of expertise relating specifically to your product's industry. To locate a trade association, return to the library and ask for the trade association publication.

There are numerous other categories that *The Idea LogBook* will not cover because its main purpose and commitment is to developing an idea into a product. Other areas such as distribution, warehousing, transportation, sales, marketing and even advertising (just to name a few) each, just like manufacturing, could require a LogBook of its own to cover their many intricacies. It will be up to the business owner to seek the needed information and education on all these functions or hire individuals who have the expertise.

TIME FOR THOUGHT

1. Why do you want to start your own business?

....

....

....

....

2. Where do you see your business in 1, 3, 5 and 10 years?

....

....

....

....

....

....

....

....

....

....

3. What areas do you feel are your strengths to operate a business?

....

....

....

....

4. What areas do you require additional expertise in operating a business?

....

....

....

5. Read at least one book on starting your own business. List any key points you'd like to reference later.

_____Name of book

....

....

....

....

....

....

NOTES

NOTES

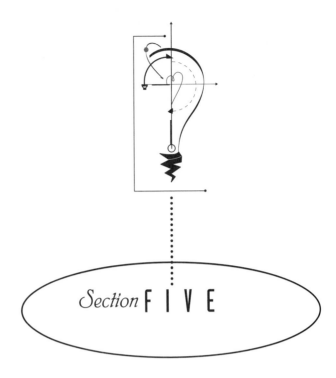

Section F I V E

In this section we will cover
financial considerations.

•

There are also some final
suggestions included to help you
decide the best way to
get your product to market.

•

You will want to make sure
that you utilize the forms
provided to track your
expenses and time.

FINANCIAL

This is the section most inventors fear (and you know what fear means) because it requires a greater amount of soul searching and creativity. Your ability to obtain finances is a direct challenge to your commitment in the development of your idea. In the face of adversity, you must choose if your idea will succumb to failure or blossom into the potential of your dream.

☞ *RULE # 12 - MONEY IS GOOD.*

You can do more good with money to improve yourself and society than you can do without it. It is not money that makes the character. It is our belief and identity about money that shapes character.

Ideas require finances to develop. The type and complexity of the product will determine the amount and depth of financing needed. There are hundreds of ideas just sitting in an undeveloped stage because they require financing to proceed. If the inventor just applied a little effort and synergy, most of their ideas could have already been on the market.

Where Does the Money Come From?

There are several avenues to choose from as to where and how you can obtain financing, and though this is not a complete list, it does provide numerous suggestions. At this stage, you know the basic costs of developing your product and your decision should be based on your financial outlook.

You

The first place to look for financing is of course, you. You have to take an inventory and determine your assets as to what (if any) amount of moneys can be diverted to developing your product. This should include savings, credit cards, stocks, bonds, etc. This is not a course on personal financing and by no means do we recommend or encourage you to go beyond all your means, it is just an exercise for you to take inventory of your finances in relation to developing an idea.

Barter

The next method for obtaining financial assistance depends on your qualifications and capabilities. It is to barter your natural or obtained skills for the skills required in developing your product. For example, let's say you possess a certain motivating or speaking skill, why not trade your services for the services you require. You do have special skills and they may be more in demand then you think. Ask if they will let you barter your time or service for their service.

Starting from this point on, it is recommended that you use a confidential, non-compete agreement with everyone with whom you discuss the product. It will be in everyone's best interest. Review "Protecting the Idea" (Chapter 4), if necessary.

Family and Friends ... Perhaps That Long Lost Uncle in Des Plaines....

The next place would be family and friends—those close relationships that trust and believe in you. If you're fortunate and have family that is so close that they will just give you the money in hopes of being repaid, you are better off than most. In reality, most families and close friends will want to be repaid either with interest or a share of the profits.

A Partner

Of course as mentioned earlier, you could bring in a partner who has the needed skills and additional finances. To give a portion of the profit and get your idea developed and into the market is better than letting it die and earning nothing.

Small Business Administration

Another choice is the Small Business Administration (SBA). They have numerous programs in existence that just could fit with your requirements including the Small Business Investment Corporations (SBIC) which financially assist small businesses with startup and growth situations. Contact the SBA at www.sba.gov or 1-800-827-5722 for further information.

As a reminder, contact SCORE through the SBA and get their input as to the latest and most popular method for obtaining financing. They are retired business executives with a wealth of information.

The greatest disadvantage with dealing with the SBA is that it is a government agency and though they can provide a wealth of benefit to you, it does take considerably more time and requires a tremendous amount of paper work.

There's Always the Bank

Commercial banks are a source for short-term loans usually with a maximum life of about 5 years. In the past, banks have been very unfriendly to small businesses and ventures. They have become a little more open and receptive to at least listening to you, especially if you choose to create a business from your product.

Venture Capital

Another source that should be considered is Venture Capital, as it is becoming a great source for funds for new business ventures. These firms usually loan money for a percentage of ownership of the business.

There exists something called *Vulture Capital*, so be sure to interview several venture capital firms and ask around for referrals so that you obtain fair financing that best fits you and your product. No one should give up seventy percent of their business.

Supplying Vendors

Other sources for generating the capital required to develop your idea are vendors from whom you obtain supplies, equipment manufacturers and distributors, insurance companies and sales finance companies. These organizations are comfortable giving loans or negotiating payments with the right opportunity. It costs nothing to ask.

If your product is completely developed and you spent time marketing and are able to secure an order for a large quantity, take that purchase order to any one of these lending institutions (including family and friends) and you will discover them to be quite receptive. A purchase order is like having guaranteed income or collateral for a loan.

Issue Stocks and Bonds

Another method for obtaining long-term financing is to establish a business or an enterprise that issues stocks and bonds (see "corporations" in Chapter 11, "Creating a Business"). The advantage would be to use other people's monetary investment into your organization to finance its operation. This may be difficult especially with only one or two products. This method has strict guidelines and regulations that need to be adhered to and it would be in your best interest to consult your attorney.

Careful thought is required as to the financial direction you choose to take and these are just a few suggestions of resources for guidance as to the best choice that fits your needs so that your idea is developed to its full potential.

Another Area To Be Aware of Is Taxes.

The government has laws stating whether and how you can or cannot write off the expenses incurred while developing an idea. In the United States, it has to do with your idea being considered a hobby and if you have made any money from it. Consult you tax accountant for the exact laws and regulations. Be sure they sign a confidential agreement if you provide details of the idea. But you should be protected under the client agreement. Ask just to be sure.

Anything purchased towards the development of your idea including photocopying, material, parts, attorney fees and work out-sourced, should be noted and tallied to assure proper accounting for credits and deductions that may be available. The following page is provided so that your records stay organized.

As you incur costs, simply indicate the date, name of store where items were purchased, a description and purpose of the items. If you out-sourced any work, be sure to also record these costs. Place the actual receipts in a separate envelope for safekeeping.

The page following the Expense Report pages, is a Timecard. It is to be used to record the length of times you worked on developing your idea, including the prototype. It is also to provide a trail of proof that you continuously worked on developing your idea so as not to have abandoned it. This along with your continuously dating any changes will aid in reassuring its protection.

Be sure to especially note the times for the prototype as the hours need to be tallied separately for determining costs.

Following the expense and timecard forms is a technique to assist you in raising funds through obtaining a short-term loan. Only you know your current financial outlook. You should only use this if it fits within your financial situation.

EXPENSE REPORT

Name of Product _____

Date	Description	Purpose	Amount
Section 1			
		Sub Total	
Section 2			
		Sub Total	
Section 3			
		Sub Total	
		TOTAL	

EXPENSE REPORT CONT.

Name of Product _____

Date	Description	Purpose	Amount
Section 1			
		Sub Total	
Section 2			
		Sub Total	
Section 3			
		Sub Total	
		TOTAL	

EXPENSE REPORT CONT.

Name of Product _____

Date	Description	Purpose	Amount
Section 1			
		Sub Total	
Section 2			
		Sub Total	
Section 3			
		Sub Total	
		TOTAL	

EXPENSE REPORT CONT.

Name of Product _____

Date	Description	Purpose	Amount
Section 1			
		Sub Total	
Section 2			
		Sub Total	
Section 3			
		Sub Total	
		TOTAL	

E X P E N S E   R E P O R T CONT.

Name of Product _____

Date	Description	Purpose	Amount
Section 1			
		Sub Total	
Section 2			
		Sub Total	
Section 3			
		Sub Total	
		TOTAL	

TIME CARD

Name of Product

Date	Time In	Time Out	Total Hrs.		Date	Time In	Time Out	Total Hrs
TOTAL					TOTAL			

TIME CARD CONT.

Name of Product

Date	Time In	Time Out	Total Hrs.		Date	Time In	Time Out	Total Hrs
TOTAL					TOTAL			

TIME CARD CONT.

Name of Product

Date	Time In	Time Out	Total Hrs.		Date	Time In	Time Out	Total Hrs
TOTAL					TOTAL			

Raise $10,000 to $50,000

This technique is to raise (some of) the money required to develop your product from five to seven "personal and business" associates—people you know and trust. The objective is to raise a personal loan from people who believe and trust in you, wanting you to succeed because of your commitment to and strength of your vision.

The greatest advantage of this technique is that it allows you to maintain control of your venture. You will not be giving away any part of your business. This is the very simple approach that has been used by some very powerful organizations as they began their business endeavors.

This first portion is to prepare the right information and tools to place in a portfolio for potential investors. Follow these easy steps.

Step 1. Complete the business plan as indicated in Chapter 11, *"Creating a Business."*

Step 2. Take a brief synopsis of the business plan and place them on two (at the very most three) pages.

Step 3. Indicate an overview of what you expect to accomplish and how it will work.

Step 4. Create a 3-4 page marketing plan, even if your goal is to license your product.

Step 5. Create a contract (you may want to obtain advice from a lawyer). Indicate your name, the loaner's name and blank spaces (lines) for the amount and date. Include the interest you feel would be valuable to the individuals loaning you the money (check with your local bank to see what the going rate is).

Also, indicate what the loan is for, its duration (six months, year), and that it is personally guaranteed and based on good faith and trust of both parties. At the end input two signature and date blocks for you and the person loaning you the money.

Fill in the blanks at the time you and the lender agree and sign the contract.

Step 6. Get business cards made with your name, phone number, fax, email and logo if you created one. Order a small quantity.

Step 7. Place this information along with a business card into a conservative binder.

Step 8. Have prototype (preferably) or drawings of your product ready to show.

Step 9. Generate a list of names of everyone you know with whom you have a personal or business relationship who may have the resources to loan.

Step 10. Prioritize the list of names.

Step 11. Make several copies of the completed report.

This second section is where you will actually sit down with individuals and request the loan.

Step 12. Begin contacting the names on the prioritized list by phone.

Step 13. Ask them if they would sit down with you to discuss a business proposition.

Step 14. If they ask why, or what's up, inform them you are raising some money and giving them the first opportunity to benefit from a very attractive offer.

Step 15. Arrange a meeting at their convenience.

Step 16. If they say no, they're not interested, thank them very much and move to the next person on your list.

Do this until you have arranged about 6 or 7 appointments.

Step 17. Meet with them precisely as arranged, sit down and hand them your business card.

Step 18. Be very straightforward and sincere and explain that you are looking to raise money from several close personal and business associates for your business venture. Explain that you are giving them the first opportunity to benefit.

Step 19. Also explain that you are going to share a concept with them that they may or may not feel makes sense and by no means, do you expect it to negatively impact the friendship.

Step 20. Explain the key points from the synopsis in the portfolio.

Step 21. Show and explain to them the prototype or drawings. You may first have them sign a confidential agreement at this junction to continue protecting your investment.

Step 22. Pull the contract out and re-explain why the need for the loan in one or two sentences, the terms you require and interest you are willing to pay.

Step 23. Sincerely explain that this is a personal loan in which you give them your word that paying them back is your highest priority.

Step 24. Restate that you are only looking for a few special people and you are okay with a "no" decision and hope that your friendship will continue to grow.

Step 25. Inform them that you are raising $_____ and ask them how much of that amount they would like to invest.

Step 26. If they agree, great! Complete the contract, inform them you will be giving them monthly status reports and reassure them they made the right decision because repaying this loan IS your highest priority.

Step 27. Leave the portfolio in their hands.

Step 28. If they do not agree, thank them for their time and inform them that you respect and accept their decision and hope the friendship continues to flourish.

Step 29. Proceed to the next person until you reach your required dollar amount.

Investors?

During this process of obtaining a loan, you may discover that some of these close associates have strong feelings toward the potential of the idea. They may ask to be able to invest in it, instead of just giving you a loan. Their reasoning is because of their strong feelings about the product's potential and the fact that they do not want to miss out on a good thing that could provide them a greater return than just the interest you agreed upon.

This is a good thing in the fact that it is an indication to you of your product's potential and success. Of course all investments do not need to be paid back if the product does not meet the financial expectations.

If this is something you might consider, review "Creating a Business" (Chapter 11) and clarify all avenues to assure everyone understands their responsibilities. Again you may want to obtain legal counseling.

TIME FOR THOUGHT

1. Where will you obtain the necessary financing for the development of your idea?

    ....
    ....
    ....
    ....
    ....
    ....

2. What other resources are available to you?

    ....
    ....
    ....
    ....
    ....
    ....

3. What other resources can you generate?

    ....
    ....
    ....
    ....
    ....
    ....

4. What skills do you possess that may be used for bartering?

    ....
    ....
    ....
    ....
    ....
    ....

5. What are the tax laws governing deductions in relation to you and your product?

    ....
    ....
    ....
    ....
    ....
    ....

FINANCIAL

NOTES

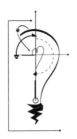

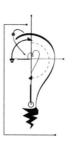

NOTES

THE FINAL SUGGESTIONS

It is now the exciting time for putting together everything in your plan and choosing which direction your efforts will take you as your product makes its way to the marketplace.

You have traveled a path that has brought you to a higher level of experience and, though it did present itself with some challenges, the overall process of this journey should have been fun. It is time to reap the rewards of your time and effort.

☞ *RULE # 13 - BE RESPONSIBLE.*

You and you alone are responsible for your own actions and decisions.

Return to "Conceptualization" (Chapter 2) and review your reasons for traveling this journey. Did you want to create a product to provide you with your sole source of income? Or, be a vehicle to develop your other product ideas? Or, allow you to create a business? Or, will it be to provide you the freedom to go after your other true dreams?

If you find you still need a little more direction, review "Now What?" (Chapter 9) and the results from the survey of your qualifications. Where did you stand in your desire, skills and qualifications?

As you review the avenues below and clearly understand the options available to determine the direction in which you and your product will travel, be certain that there is no wrong choice except not making a decision. Your decision should be made based on you and your purposes alone. The desires and dreams that are within you will help you determine this decision. Whatever you decide is the right choice.

Three Ways You Can Go From Here

There are three basic avenues to consider in determining the best direction to bring your product to market. Each one provides its own unique advantages and disadvantages. They are not listed in any order of preference or recommendation. No one avenue is better or worse than the other. You are looking to make your decision based on what would be best for you and your idea. It all depends on what you choose.

Create a Business

The first avenue is to create a business to develop, manufacture, distribute and market the product. This avenue requires the most time and effort as you will be totally responsible for overseeing the complete process. The greatest advantage with this avenue is that it allows you complete control over the product, creates employment for you (and others), fulfills the entrepreneurial dream of business ownership and provides the opportunity of generating the greatest amount of income.

The greatest disadvantage to creating your own business is that you are entering a market with only one product and products have life cycles that seem to be growing shorter. It might challenge and stretch the boundaries of every skill and qualification that you possess, which actually should be a good thing.

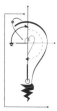

Outsourcing

The second avenue is to outsource different portions of the process while you handle some of the processes yourself. For example, if you wanted to create a business and promote and market the product but, because of a lack of resources, (skills, finances or even choice), you could hire a manufacturer to produce the product. Or, if you choose to be the manufacturer, you can outsource or subcontract a company to conduct the marketing of your product.

The advantage of outsourcing some of the processes is that it provides you with strength in an area where you might have lacked skills or expertise. This option keeps you in complete control of the product and fulfills the entrepreneurial dream of business ownership. It also allows you the advantage of being able to focus more of your time and energy on your areas of expertise or other important dreams.

The greatest disadvantage is that the product is partially in the hands of someone whom you will have to keep a close business eye on to assure they are performing to your standards. Be sure to create a great synergy between you and the out-sourced company you outsource with. Again, it's advisable to retain the services of a business attorney to assure everyone understands their responsibilities and requirements.

Licensing

The third avenue is to license out the idea to a company that will take your product and produce and market it, paying you a royalty. See "Licensing" (Chapter 10) for more details.

The advantage of licensing is that it puts your product in the hands of experts who know and understand the industry and have the resources and skills to develop, manufacture, distribute and market your product.

The greatest disadvantage is that it *does* require time and effort on your part to locate and create a match for the product to assure it is going to be properly developed and marketed. Giving up control (partial or complete) of your product should not be considered a disadvantage if you're looking at this option. It shouldn't even be playing a role if your best interest lies in taking your idea and turning it into a product that gets to the market. The feeling of loosing control is a fear, which is a behavior we already discussed, remember? Focus your attention on the *outcome* of seeing your product on the shelves of your local store.

TIME FOR THOUGHT

1. What are your overall dreams for this product?

....
....
....
....

2. Why would you choose to create a business to manufacture and market your product?

....
....
....
....
....
....

3. What would the advantages be for you to create a business and outsource some of the processing stages?

....
....
....
....
....
....
....

4. What would be the advantages for you to license your product to a company?

....
....
....
....
....

5. Which avenue appeals to you most? Why?

....
....
....
....
....
....

NOTES

THE FINAL SUGGESTIONS

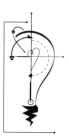

NOTES

THE FINAL SUGGESTIONS

CONCLUSION

Having worked your way through *The Idea LogBook*, you now have more than enough information to take an idea and conduct the necessary research and development to determine if it is a viable product. You have taken the required action and with minimal time and effort, you have saved hundreds (possibly thousands) of dollars while enjoying the process of an exciting journey.

You have accomplished more than most people would have and are now in a better position to develop your idea into a product and then bring this product to the market. You have learned valuable information that will allow your mind to be more open to receive other ideas and the process of developing your next idea now has been made more simple.

Now go out, get another idea, order another *Idea Logbook* and have some more fun. Your next adventure and royalties await!

Feel free to contact us at The Inventors' Place. We can be reached at P.O. Box 4589, Oceanside, CA 92052 or www.Inventorsplace.com.

The Inventors' Place is implementing systems and programs to assist you in developing your idea into a product and to get it to the market.

Our website is designed to be your "inventing resource center" for product development as we are constantly adding sources of information for you to access to ease and simplify the development of your idea.

Also, The Inventors' Place has created a vehicle which will allow you to immediately begin selling your developed product and, of course, purchase other unique inventions.

We will continue to ease the idea development process as we currently have numerous programs under consideration including one to provide you with assistance and support (should you require) to develop your idea into a solid invention.

In conclusion, *The Idea LogBook* is dedicated to the
"Creativity of Inspiration"
and the belief that individual inventors
are the true innovators of society.
We encourage everyone to dream and create.

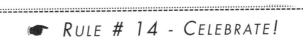

☞ *RULE # 14 - CELEBRATE!*

You Deserve It, You Deserve It.

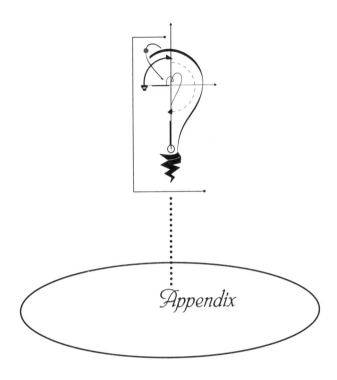

Appendix

The following pages are
included to provide you with
additional space to keep track of
your growing idea.

•

Remember to continue to
document your idea
throughout its
evolution.

THE TOP 14 RULES FOR INSPIRATION

These rules are provided to assist in your journey so that it broadens your creativity and increases your inspirational outlook.

Rule # 1 -   Have Fun

Rule # 2 -   Right Action = Right Results

Rule # 3 -   Be Open

Rule # 4 -   It's Easy

Rule # 5 -   Focus on Profits and Outcome

Rule # 6 -   Faith in Self

Rule # 7 -   Willingness to Learn and Take Risks

Rule # 8 -   Be Aware

Rule # 9 -   Acceptance

Rule # 10 - Be Humble not Reckless and Arrogant

Rule # 11 - Reject the Status Quo; Strive for Something Greater

Rule # 12 - Money Is Good

Rule # 13 - Be Responsible

Rule # 14 - Celebrate, You Deserve It

PRODUCT DIAGRAM

Product Name: _____   Date: _____

Witnessed and understood by: _____   Date: _____

PRODUCT DIAGRAM

Product Name: _____   Date: _____

Witnessed and understood by: _____   Date: _____

P R O D U C T   D I A G R A M

Product Name: _____   Date: _____

Witnessed and understood by: _____   Date: _____

PRODUCT DIAGRAM

Product Name: _____   Date: _____

Witnessed and understood by: _____   Date: _____

PRODUCT DIAGRAM

Product Name: _____   Date: _____

Witnessed and understood by: _____   Date: _____

P R O D U C T D I A G R A M

Product Name: _____   Date: _____

Witnessed and understood by: _____   Date: _____

PRODUCT DIAGRAM

Product Name: _____   Date: _____

Witnessed and understood by: _____   Date: _____

P R O D U C T   D I A G R A M

Product Name: _____   Date: _____

Witnessed and understood by: _____   Date: _____

PRODUCT DIAGRAM

Product Name: _____   Date: _____

Witnessed and understood by: _____   Date: _____

PRODUCT DIAGRAM

Product Name: _____   Date: _____

Witnessed and understood by: _____   Date: _____

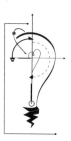

NOTES

NOTES

NOTES

NOTES

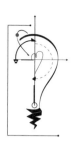

NOTES